Galois' Dream:

Group Theory
and
Differential Equations

For Kyoko

Michio Kuga

Galois' Dream:
Group Theory and Differential Equations

Susan Addington
Motohico Mulase
Translators

Birkhäuser
Boston · Basel · Berlin

Translator:
Susan Addington
Department of Mathematics
California State University
San Bernardino, CA 92407
USA

Translator:
Motohico Mulase
Department of Mathematics
University of California
Davis, CA 95616
USA

Library of Congress Cataloging-in-Publication Data

Kuga, Michio, 1928-1990.
  [Garoa no yume. English]
   Galois' dream: group theory and differential equations / Michio
  Kuga ; translated by Susan Addington and Motohico Mulase.
      p.   cm.
   Includes bibliographical references and index.
    ISBN 0-8176-3688-9 (alk. paper). -- ISBN 3-7643-3688-9 (alk.
  paper)
     1. Galois theory.   2. Differential equations.   3. Monodromy
  groups. I. Title.
  QA174.2.K8413   1993   92-41486
  512'.3--dc20   CIP

Printed on acid-free paper.

ISBN 0-8176-3688-9
ISBN 3-7643-3688-9

Typeset in TeX by by Susan Addington.
Printed and bound by Quinn-Woodbine, Woodbine, NJ.
Printed in the U.S.A.

9 8 7 6 5 4 3

Contents

Preface vii

Pre–Mathematics

0th Week	No prerequisites	3
1st Week	Sets and Maps	5
2nd Week	Equivalence Classes	19
3rd Week	The Story of Free Groups	23

Heave Ho! (Pull it Tight)

4th Week	Fundamental Groups of Surfaces	33
5th Week	Fundamental Groups	38
6th Week	Examples of Fundamental Groups	45
7th Week	Examples of Fundamental Groups, continued .	49

Men Who Don't Realize
That Their Wives Have Been Interchanged

8th Week	Coverings	53
9th Week	Covering Surfaces and Fundamental Groups .	59
10th Week	Covering Surfaces and Fundamental Groups, continued	61
11th Week	The Group of Covering Transformations . . .	70

Everyone Has a Tail

12th Week	The Universal Covering Space	77
13th Week	The Correspondence Between Coverings of $(D; O)$ and Subgroups of $\pi_1(D; O)$	83

Seeing Galois Theory

14th Week	Continuous Functions on Covering Surfaces .	89
15th Week	Function Theory on Covering Surfaces	93

Solvable or Not?

16th Week	Differential Equations	105
17th Week	Elementary Methods of Solving Differential Equations	109
18th Week	Regular Singularities	114
19th Week	Differential Equations of Fuchsian Type . . .	129
	References	141
	Notation	143
	Index	147

Preface

They came back to say that they all fell into the crevasse.

These are lecture notes for a course I gave at the University of Tokyo a few years ago. The forty students who attended were first year undergraduates. Among these, about twenty five completed the course. Rather than giving a simplified presentation of the lectures here, I will reproduce the complete set of lectures, preserving the original style faithfully. The title of the course was "Group Theory and Differential Equations."

What should one expect from the title "Group Theory and Differential Equations"?

It is well known that group theory, at its inception, was deeply connected to algebraic equations. The success of the theory of Galois groups in algebraic equations led to the hope that similar group-theoretic methods could be a powerful arsenal to attack the problems of differential equations. In fact, Sophus Lie devoted his life to this problem, inventing the theory of Lie groups as a weapon. Although this brought him lasting fame, he was unable to make substantial progress in his original aim of a group-theoretic study of differential equations. Picard and Vessiot inherited the problem, and later it was studied by Ritt and Kolchin from the algebraic point of view. This work has been very successful for some special types of differential equations, including linear ordinary differential equations, but it seemed far from satisfactory compared with the theory Lie envisioned. In order to follow Lie's original program, it would be necessary to use the ideas and techniques of modern analysis, such as the infinite dimensional Lie groups of E. Cartan and Kuranishi and the cohomology theory of D. C. Spencer. [More recently (1990), Kimura, Hattori, Sato, Kashiwara, and Yoshida have also contributed to the subject.]

By the way, these lectures are not an introduction to these important theories (although the subjects are not completely unrelated).

My lectures are on Fuchsian differential equations and their monodromy groups. Riemann also tried a group-theoretic treatment of linear ordinary differential equations. He used discontinuous groups rather than the continuous groups that Lie used later. These discontinuous groups are defined purely topologically, and are used to understand the way the domains of the differential equations are connected. These groups are called fundamental groups or monodromy groups. They also represent the multivalued nature of the solutions of the differential equation. In fact, the fundamental group Γ is contained in the continuous group G considered by Lie, Picard, and Vessiot. (We call it the Picard-Vessiot group later on.) However, since $\Gamma \neq G$, Γ is not as powerful as the Picard-Vessiot group in solving differential equations. But if we restrict ourselves to only Fuchsian type differential equations, then the solutions are completely characterized by the monodromy group Γ. This topic will be the main subject of this book.

The main importance of the monodromy group, however, is not in the theory of solvability of differential equations. Rather, Γ shows its true power when the differential equation cannot be solved by elementary methods. The monodromy group connects the theory of Fuchsian differential equations of special types with the theory of automorphic functions and enables us to investigate the deep analytic and algebraic structure of the solutions. Riemann treated the case with three singularities—that is, the case in which the solutions are essentially described by hypergeometric functions. Riemann's P-function theory is nothing but this. This approach was continued later by Fuchs and Poincaré. Poincaré studied the subject with relation to automorphic functions.

Some might criticize my choice of such a dusty old-fashioned topic. It seems puny compared with the healthy, gushing stream of modern analysis.

Nevertheless, I have chosen this topic for pedagogical reasons. Namely,

(i) It can serve as an introduction to algebraic systems (groups), topology, and analysis (function theory).

(ii) Only elementary knowledge of these three subjects is needed, but all three are necessary. This theory is the most primitive example that the intersection of many different disciplines is the most interesting mathematics. It can help convince students that mathematics is a unity; it also gives a wider perspective of mathematics.

(iii) It gives a geometric understanding of Galois theory.

(iv) There are many challenging problems which are not too difficult for student exercises.

(v) Above all, I cannot forget the old dogma that the most interesting aspect of analysis is its algebraic structure.

It may be true that there are no more good research problems left in the theory of Fuchsian differential equations, but the situation is more difficult in several variables. There seem to be many new directions to explore here. For example, it doesn't seem possible to reduce Appell's theory of hypergeometric functions in two variables to the theory of symmetric spaces. I wonder if it is possible to reduce the case of discrete monodromy groups to the moduli theory of algebraic varieties. I think that Kähler studied such problems in the 1930's, but they seem to be untouched since then. Is it possible to understand Leray's recent work on the Cauchy problem in partial differential equations as preliminary to a group-theoretic study of differential equations?

[There is much recent work on hypergeometric functions in several variables by mathematicians including Deligne, Manin, and Gelfand.]

Actually, I believe that the lecture series was quite successful. Some of the students became math majors, and went on to become good mathematicians. And they came back to say that they all fell into the crevasse.

September, 1967

Translator's note: When this book was first published, it was a best seller; students would carry it around to be hip, whether or not they could read it. However, it got Kuga in trouble with the mathematical establishment, as it was considered undignified for a professor at Tokyo University to publish a book with cartoons and funny examples.

More recently Kuga was asked to write another book in the same style, since Galois' Dream was so successful in attracting students to mathematics. He declined, saying, "I was only the Pied Piper."

The Japanese edition included two more chapters: The Infinityth Week and The Infinity + 1st Week. As the numbers suggest, they were at a much more advanced level than the rest of the book. Unfortunately, these chapters contained fundamental errors and needed to be completely rewritten. We decided to omit them from this edition.

Feminist readers are asked not to take umbrage at some of Kuga's examples; keep in mind that the book was written in another era and in another culture. Kuga had a deep respect (and admiration) for women.

This translation was a joint effort; it could not have been done with only one of us. Mulase translated from the Japanese; then I refined the text into idiomatic English. We also consulted a preliminary translation that Tadatoshi Akiba did in 1975.

I would like to thank the many people and institutions who provided assistance for this project: the Paul and Gabriella Rosenbaum Foundation, for providing some financial support; California State University at San Bernardino, for help with computer software and hardware, especially the staff of the Computer Center and Audio-Visual Departments; Lillian Kinkade, who TeXed the entire first draft of the book, learning TeX especially for the occasion; the many people whose laser printers I borrowed, especially Dave and Shari Stockwell; Madge Goldman, for serving as a liaison between translators, editors, and others; and Han Sah, for mathematical and editorial advice.

Finally, I wish to acknowledge a profound debt to Michio Kuga, my mathematical parent, who died February 13, 1990.

Susan Addington
April, 1994.

Pre-Mathematics

The 0th Week:
No Prerequisites

The title of this series of lectures is "Group Theory and Differential Equations". The contents are briefly explained in the preface. The 0th week is a summary of the preliminary meeting of the course. (In the lecture, I outlined everything in the course, and left the students in a fog.)

Let me describe the prerequisites you need to follow these lectures.

(0) There is no prerequisite for the first two weeks. Later you will need some group theory, introductory topology, theory of functions, and differential equations. Do not be intimidated at this point. As you will see, you won't need more than the introductory parts of these subjects.

(1) By the third week (you still have two full weeks to study!) you will need some group theory. Learn the definition of a group and some fundamental results. It is enough to understand the concepts of group, subgroup, normal subgroup, homomorphism, and isomorphism. Nothing beyond this will be used. While this material can be found in any introductory book on abstract algebra (also called modern algebra), concise treatments are given in Pontryagin [11] and van der Waerden [15].

(2) If you do not know the language of topology, find out what the terms "neighborhood", "open set", "continuous map", and "homeomorphism" mean by the 12th week. You can find these definitions in any textbook on point-set topology (also known as general topology). Singer and Thorpe [12] covers these topics in the first few pages, and [11] also contains a brief but complete summary.

(3) Starting in the 15th week, you will need some knowledge of the theory of functions of a complex variable. You need to know about holomorphic functions, the Cauchy-Riemann equations, contour integrals, Cauchy's Theorem (*i.e.*, $\int_C f(z)dz = 0$), Morera's Theorem, Cauchy's integration formula, power series, meromorphic functions, poles, and Laurent series. (There is a rapid review of these topics in Week 15.) This material can be found in any textbook on complex analysis; Ahlfors [1] is a standard one.

(4) Some facts about linear differential equations in a complex variable will be needed starting in the 16th week (especially the theorem on existence of solutions). By the 17th week, you will need to know about differential equations of Fuchs type. A good reference is Birkhoff and Rota [4]; an older book is Ince [8].

(5) From the 16th week on, some knowledge of the simplest notions of linear algebra will be needed: linear transformations and their matrix expressions. (Only two-dimensional vector spaces are used.) See, for example, Curtis [5], or any book on linear algebra. Also look up the definition of a linear representation of a group in a graduate-level algebra book. It can be found in [15], for example.

The Schaum Outline Series has inexpensive paperbacks which generally present subjects in a straightforward way; the series has books on topology, algebra, linear algebra, and complex analysis. See the references for titles and authors.

The First Week:
Sets and Maps

This week I will explain the concepts of "set" and "map", which are fundamental in mathematics. I am sure that many of you know these terms already, so my explanation will be just an appetizer before the main course.

Sets.

We call any collection of objects that is clearly determined a set. Each object in this collection (= set) is called an element. We use capital letters $\mathcal{M}$, $\mathcal{N}$, etc., to denote various sets, and small letters x, y, etc., to denote elements of these sets. When x is an element of (say) $\mathcal{M}$, we sometimes use such expressions as "x belongs to $\mathcal{M}$" and "x is contained in $\mathcal{M}$". In mathematical notation,

$$\mathcal{M} \ni x \quad \text{or} \quad x \in \mathcal{M}.$$

If y is not an element of $\mathcal{M}$, we use the notations

$$\mathcal{M} \not\ni y \quad \text{or} \quad y \notin \mathcal{M}.$$

Example 1. Consider all the natural numbers. Of course, you know what they are: $1, 2, 3, 4, 5, \ldots$. These numbers are obtained by starting from 1 and adding 1 some number of times. Although there are infinitely many natural numbers, we will consider them all at once. This collection of all natural numbers is a set. We use the letter **N** to represent this set. That is,

$$\mathbf{N} = \{1, 2, 3, 4, 5, \ldots\}$$

All the natural numbers belong to **N**, and nothing but natural numbers belongs to **N**. The number 1966 is a natural number; so $1966 \in \mathbf{N}$. But a negative number -7031 is not a natural number; so $-7031 \notin \mathbf{N}$.

Example 2. Now let's consider all the integers. An integer is any of

1) Natural numbers: $1, 2, 3, 4, 5, \ldots$

2) Natural numbers numbers with a negative sign: $-1, -2, -3, -4, -5, \ldots$ and

3) 0.

In mathematics, the letter **Z** is used to denote the set of integers.

$$-7031 \in \mathbf{Z}, \text{ because } -7031 \text{ is an integer.}$$

$$\frac{355}{113} \notin \mathbf{Z}, \text{ because } \frac{355}{113} \text{ is not an integer.}$$

Example 3. Next, consider all the rational numbers. A rational number is a number that can be written as the quotient $\frac{m}{n}$ of two integers m and n. ($n \neq 0$, of course!) The letter $\mathbf{Q}$ denotes the set of all rational numbers.

$$-\frac{355}{113} \in \mathbf{Q}, \text{ because } -\frac{355}{113} \text{ is a rational number.}$$

$$\sqrt{2} \notin \mathbf{Q}, \text{ because } \sqrt{2} \text{ is not a rational number.}$$

Example 4. The totality of real numbers forms a set, which we denote by $\mathbf{R}$. $\sqrt{2} \in \mathbf{R}$, $1 + 2\sqrt{-1} \notin \mathbf{R}$.

Example 5. The set of all complex numbers is denoted by $\mathbf{C}$. A complex number is of the form $x + \sqrt{-1}y$ (x and y are real numbers.)

Example 6. The set of all human beings alive at this precise moment ($=$ October 31, 1960, 1:32 + 17.8322... seconds.)

Example 7. The set of all students in this room right now.

Example 8. The set of all prime numbers smaller than 100. This is the set

$$\{2, 3, 5, 7, 11, 13, 17, 19, 23, 29, 31, 37, 41, 43, 47, 53, 59, 61, 67, 71, 73, 79, 83, 89, 97\}$$

consisting of 25 elements.

Example 9. Consider a plane. For example, we can imagine stretching the blackboard in this room infinitely in all possible directions. Let's call this infinite blackboard π. There are infinitely many points on this plane π. (So saying, the instructor marks many points on the board.) All of these are points of π. Therefore, the plane π can be considered as a set (that is, the set consisting of the points of π). If the point P is on π, then we write $P \in \pi$. If the point R is not on π, then we write $R \notin \pi$.

Example 10. Similarly, consider a circle C on π. (The instructor proudly draws an almost perfect circle on the board.) This circle C is also a collection of points. Therefore, it is also a set. In Figure 1.1,

$$P \in C, \quad \text{and} \quad Q \notin C.$$

A set consisting of (geometric) points is called a point set. The sets in Examples 9, 10, and 11 are, therefore, point sets.

Example 11. Let's consider an arbitrary shape on π. (The instructor tries to draw O-ba Q.*) Let's call the shaded part D. Since D consists of the points that are part of D, D is another set.

---

* a well-known Japanese cartoon ghost

Figure 1.1.

Warning: It is not necessary to consider shapes which are a single piece. For example, (the instructor draws a figure next to D, and calls it P-ko.**) You can consider these two figures at once and call them a set F.

Figure 1.2.

A point set that consists of more than one piece, such as F, is called disconnected. We will discuss this later. (The 4th week.) For example, the set of points on the land on this globe is a disconnected point set.

Figure 1.3.

The sets in Examples 6, 7, and 8 have only finite numbers of elements. The set in Example 8 consists of 25 elements, the set of Example 6 has approximately 3

** O-ba Q's sister

billion members, and the set in Example 7 (er . . . let me count: one, two, three, . . .) consists of _____ members.

Sets such as these which have finite numbers of elements are called finite sets. On the other hand, sets with infinite numbers of members or elements are called infinite sets, as in Examples 1, 2, 3, 4, 5, 9, 10, and 11. If a finite set $\mathcal{M}$ has m elements, we express this by

$$|\mathcal{M}| = m.$$

That is, $|\mathcal{M}|$ stands for the number of elements the finite set $\mathcal{M}$ contains. If $\mathcal{M}$ is an infinite set, we write

$$|\mathcal{M}| = \infty.$$

Example 12. If $\mathcal{M}$ is the set of all prime numbers less than 10, then $\mathcal{M} = \{2, 3, 5, 7\}$. Therefore $|\mathcal{M}| = 4$.

Example 13. If $\mathcal{M}$ is the set of prime numbers between 6 and 9, then $\mathcal{M} = \{7\}$. Therefore $|\mathcal{M}| = 1$.

We defined a set to be a collection of objects. In English, a collection usually means a set containing more than one thing. However, we will find it easier in describing various mathematical situations to interpret a "collection of objects" as a collection of one object, as a special case. By this convention we allow $\mathcal{M}$ in Example 13 to be considered as a set, even though $|\mathcal{M}| = 1$.

But even more bizarre, we will consider a set $\mathcal{M}$ that satisfies $|\mathcal{M}| = 0$. As long as we stick to the original description of a set being a collection of objects, it is invalid to say that this definition makes a collection of no objects a set. But in mathematics (as in many other places) we turn a blind eye to the law. So we insist that the collection without any elements is a set, and call it the empty set. We use the symbol $\emptyset$ to denote it. (In other words, we introduce the new concept of the empty set.)

Mathematese

If you learn a few words of mathematese now, you will be rewarded later. The first word is the symbol $\forall$. This is a mathematese term meaning "all". You could also interpret this to mean "any". Actually, these two words, all and any, have logically equivalent meanings. For example, the statement "All women are stupid" (Excuse me!) and the statement "Any woman is stupid" have the same meaning, don't they? (Of course, whether or not the statements are true is a different question.)

The symbol $\forall$ was invented by upending the letter A to emphasize that it stands for the words all and any.

The symbol $\exists$ is another word in mathematese meaning "there exist(s)". This is a reversed E, which stands for the word exist.

The symbol $\neg$ represents negation (i.e., No . . . or Not . . ..)

The symbol $|$ is used often instead of "such that" or "satisfying".

The brackets {...} represent a set whose elements are listed or described between the brackets. For example,

$$\{ \text{ O-ba Q, O-ba P, Dodompa* } \}$$

denotes the set that has O-ba Q, O-ba P, and Dodompa as its elements.

Pocket Mathematese–English Dictionary	
Mathematese	English
∃	exist
¬	negation
|	such that
{...}	the set with ... as its elements

Examples of Usage

(1) $\{x \in \mathbf{R} \mid x^2 < 1\}$ is the set of all real numbers x such that x^2 is strictly less than 1.

$$-1 \qquad 0 \qquad 1$$

Figure 1.4.

In the figure this set lies between -1 and 1.

(2) $\{x \in \mathbf{R} \mid \exists y \in \mathbf{R} \mid \sin y = x\}$ is the set of all real numbers x such that there exists a real number y satisfying $\sin y = x$. Therefore, this set is equal to $\{x \in \mathbf{R} \mid |x| \leq 1\}$.

Subsets

Suppose that $\mathcal{M}$ and $\mathcal{N}$ are two sets. If all the elements of $\mathcal{N}$ are elements of $\mathcal{M}$, we call $\mathcal{N}$ a subset of $\mathcal{M}$. In notation, $\mathcal{N} \subset \mathcal{M}$ or $\mathcal{M} \supset \mathcal{N}$.

Example. A natural number is a positive integer. Therefore, any natural number is also an integer. So $\mathbf{N} \subset \mathbf{Z}$. Similarly, $\mathbf{Z} \subset \mathbf{Q}$: because any integer n can be written as $\frac{n}{1}$, it is a rational number. Also, $\mathbf{Q} \subset \mathbf{R}$ and $\mathbf{R} \subset \mathbf{C}$. In summary, $\mathbf{N} \subset \mathbf{Z} \subset \mathbf{Q} \subset \mathbf{R} \subset \mathbf{C}$.

When you have sets $\mathcal{M}$, $\mathcal{N}$, and $\mathcal{L}$, if $\mathcal{M} \supset \mathcal{N}$ and $\mathcal{N} \supset \mathcal{L}$, then $\mathcal{M} \supset \mathcal{L}$.

According to our definition of a subset, the set $\mathcal{M}$ is its own subset; i.e., we have $\mathcal{M} \subset \mathcal{M}$. Now suppose that $\mathcal{N}$ is a subset of $\mathcal{M}$, and furthermore suppose

---

* standard Japanese nonsense word

that $\mathcal{N}$ is not $\mathcal{M}$ itself; i.e., $\mathcal{N} \subset \mathcal{M}$ and $\mathcal{N} \neq \mathcal{M}$. We call $\mathcal{N}$ a proper subset of $\mathcal{M}$. When $\mathcal{N}$ is a proper subset of $\mathcal{M}$, and we want to emphasize this fact, we write

$$\mathcal{M} \underset{\neq}{\supset} \mathcal{N} \quad \text{or} \quad \mathcal{N} \underset{\neq}{\subset} \mathcal{M}$$

Convention: We always consider the empty set $\emptyset$ to be a subset of any set. In other words, for any set $\mathcal{M}$, $\mathcal{M} \supset \emptyset$.

Exercise. Let $\mathcal{M} = \{a, b, c\}$ be a finite set consisting of three elements a, b, and c. Find the total number of subsets of $\mathcal{M}$.

(Solution):

(0) $\mathcal{M}$ is a subset of $\mathcal{M}$.

(1) The sets consisting of two elements

$$\{a, b\}, \{b, c\}, \{c, a\}$$

are subsets of $\mathcal{M}$.

(2) The sets consisting of one element

$$\{a\}, \{b\}, \{c\}$$

are subsets of $\mathcal{M}$.

(3) The empty set $\emptyset$ is a subset of $\mathcal{M}$ by our convention.

Therefore $\mathcal{M}$ has the eight subsets listed above.

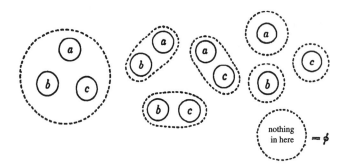

Figure 1.5.

Exercise. Let $\mathcal{M}$ be a finite set with m elements. Find the number of subsets of $\mathcal{M}$.

(Answer): 2^m. Prove this yourself.

Set Operations

We choose one set $\mathcal{M}$ and investigate the relations among its subsets $A, B, C, ...$

When A and B are two subsets of $\mathcal{M}$, we want to consider the collection of elements of $\mathcal{M}$ that belong to at least one of A or B. In mathematical notation, consider

$$\{x \in \mathcal{M} \mid x \in (\text{at least one of } A \text{ or } B)\}.$$

This is certainly a subset, and we denote it by $A \cup B$. The subset $A \cup B$ is called the *union* of A and B.

Figure 1.6.

In other words, $A \cup B = \{x \in \mathcal{M} \mid x \in (\text{at least one of } A \text{ or } B)\}$. $A \cup B$ can be represented by the shaded shape in Figure 1.6. If there are three subsets, A, B, and C of $\mathcal{M}$, we can still define the union $A \cup B \cup C$ by

$$\{x \in \mathcal{M} \mid x \in (\text{at least one of } A, \ B, \text{ or } C)\}.$$

More generally, if you have subsets $A_1, A_2, A_3, ...$ (a finite or infinite number of sets), you can still define the union

$$A_1 \cup A_2 \cup A_3 \cup ... = \bigcup_i A_i$$
$$= \{x \in \mathcal{M} \mid x \in (\text{at least one of } A_1, A_2, A_3, ...)\}$$
$$= \{x \in \mathcal{M} \mid \exists i \mid x \in A_i\}.$$

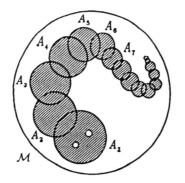

Figure 1.7.

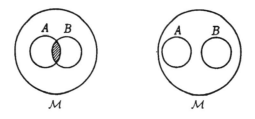

Figure 1.8.

Next consider the elements of $\mathcal{M}$ that belong to both A and B. This set is called the *intersection* of A and B, and is denoted by $A \cap B$.

$$A \cap B = \{x \in \mathcal{M} \mid x \in A \text{ and } x \in B)\}.$$

See the first part of Figure 1.8 for a pictorial explanation.

If A and B are separated, as in the second part of Figure 1.8, then $A \cap B = \emptyset$. Also,

$$A \cap B \cap C = \{x \in \mathcal{M} \mid x \in A, x \in B, \text{ and } x \in C)\}.$$

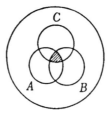

Figure 1.9.

The intersection of a finite or infinite number of subsets $A_1, A_2, A_3,...$ can be defined similarly (and is illustrated in Figure 1.10):

$$A_1 \cap A_2 \cap A_3 \cap ... = \bigcap_i A_i$$
$$= \{x \in \mathcal{M} \mid x \in (\text{all of } A_1, A_2, A_3, ...)\}$$
$$= \{x \in \mathcal{M} \mid \forall i, x \in A_i\}.$$

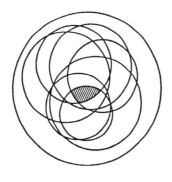

Figure 1.10.

Formula 1.1. If A, B, and C are subsets of $\mathcal{M}$, then

$$A \cup (B \cap C) = (A \cup B) \cap (A \cup C)$$
$$A \cap (B \cup C) = (A \cap B) \cup (A \cap C)$$

These two identities are called the distributive laws.

Prove these by looking at the following figures.

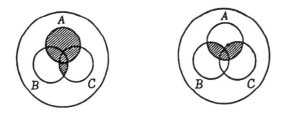

Figure 1.11.

Maps.

Let $\mathcal{M}$ and $\mathcal{N}$ be two sets. A rule that assigns to each element m of $\mathcal{M}$ an element n of $\mathcal{N}$ is called a *map from $\mathcal{M}$ to $\mathcal{N}$*. We use letters f, g, φ, ψ, etc. to denote maps. The element of $\mathcal{N}$ that f assigns to the element x of $\mathcal{M}$ is denoted by $f(x)$. We call $f(x)$ the *image* of x.

Example 15. When $\mathcal{M} = \mathcal{N} = \mathbf{R}$, a map from $\mathbf{R}$ to $\mathbf{R}$ is a "function". You may recall that it is convenient to express a function by its graph. Here are a few functions and their graphs. Let f_1 be the function that assigns $2x - 1$ to a real number x. We also use the notation $f_1 : \mathbf{R} \ni x \mapsto 2x - 1$ to describe the function.

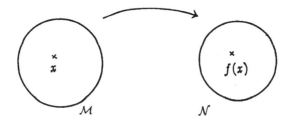

Figure 1.12.

Similarly, let f_2 be the function that assigns $x^2 + 1$ to a real x, and f_3 be the function that assigns e^x to each real x. In other words,

$$f_2 : \mathbf{R} \ni x \mapsto x^2 + 1; \text{ or } f_2(x) = x^2 + 1$$

$$f_3 : \mathbf{R} \ni x \mapsto e^x; \text{ or } f_3(x) = e^x$$

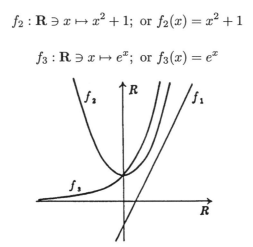

Figure 1.13.

Example 16. Let π be a plane, and l be a line on π. Let $proj_l$ be the map that assigns to each point P on π its orthogonal projection P' on l. This is a map from π to l.

$$proj_l : \pi \to l$$
$$P \mapsto \text{ (orthogonal projection of } P \text{ on } l)$$

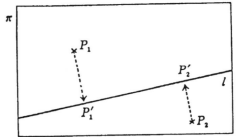

Figure 1.14.

Example 17. Let $\mathcal{M}$ and $\mathcal{N}$ be two parallel planes as in the next figure, and O be a point that is not on $\mathcal{M}$ or $\mathcal{N}$. Let's call $\mathcal{M}$ the transparency and $\mathcal{N}$ the screen, and denote by φ the map that assigns to each point P of $\mathcal{M}$ the intersection point P' of the line $\overline{OP}$ and $\mathcal{N}$.

$$\varphi : \mathcal{M} \to \mathcal{N}$$
$$P \mapsto P' = \overline{OP} \cap \mathcal{N}$$

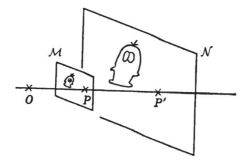

Figure 1.15.

Example 18. Let $\mathcal{N}$ be the set of all human beings ever born (including those who are no longer alive) starting from Adam and Eve. We define $\mathcal{M}$ to be the set obtained by omitting Adam and Eve from $\mathcal{N}$: $\mathcal{M} = \mathcal{N} - \{\text{Adam, Eve}\}$. Let us denote by μ the map that maps each element x (Mr. x or Ms./Miss/Mrs. x) of $\mathcal{M}$ to his or her mother.

$$\mu : \mathcal{M} \ni x \mapsto \text{ mother of } x \in \mathcal{N}.$$

Similarly, φ assigns the father to x:

$$\varphi : \mathcal{M} \ni x \mapsto \text{ father of } x \in \mathcal{N}.$$

Exercise. When $\mathcal{M}$ and $\mathcal{N}$ are finite sets with $|\mathcal{M}| = m$ and $|\mathcal{N}| = n$, find the total number of maps from $\mathcal{M}$ to $\mathcal{N}$. Before giving the answer, let's do some experiments. Let $m = 3$ and $n = 2$, and write $\mathcal{M} = \{x, y, z\}$ and $\mathcal{N} = \{a, b\}$. We use the arrow representation of functions.

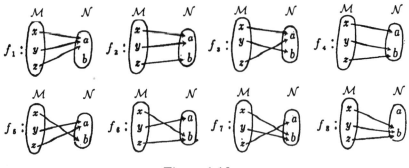

Figure 1.16.

As you see, there are 8 maps. In general, there are n^m maps. The proof is left to the reader.

Let f be a map from $\mathcal{M}$ to $\mathcal{N}$, i.e. $f : \mathcal{M} \to \mathcal{N}$. The range of $f(x)$ as x moves through all of $\mathcal{M}$ is denoted by $f(\mathcal{M})$. This is a subset of $\mathcal{N}$. In notation,

$$f(\mathcal{M}) = \{f(x) \mid x \in \mathcal{M}\}.$$

If you prefer a bit more rigor,

$$f(\mathcal{M}) = \{y \in \mathcal{N} \mid \exists x \in \mathcal{M} \mid f(x) = y\}.$$

In other words, $f(\mathcal{M})$ is the set of elements y of $\mathcal{N}$ such that there exists an element x of $\mathcal{M}$ such that $f(x) = y$. We call $f(\mathcal{M})$ the *f-image of* $\mathcal{M}$.

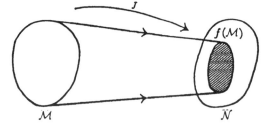

Figure 1.17.

In Example 15, $f_1(\mathbf{R}) = \mathbf{R}$, $f_2(\mathbf{R}) = [1, \infty)$, $f_3(\mathbf{R}) = (0, \infty)$. In Example 16, $proj_l(\pi) = l$.

If it so happens that $f(\mathcal{M}) = \mathcal{N}$, we say that f is a *surjection*, or f is an *onto* map from $\mathcal{M}$ to $\mathcal{N}$, or f is *surjective*. The map f_1 of Example 15, $proj_l$ of Example 16, and φ of Example 17 are surjective. If the fact that $x \neq y$ $(x, y \in \mathcal{M})$ guarantees that $f(x) \neq f(y)$, f is an *injection*, or f is *injective*. We also say that f is *one-to-one*. The functions f_1 and f_3 of Example 15 and the map φ of Example 17 are injective. The function f_2 of Example 15 is not injective. Neither is the map $proj_l$ of Example 16.

The function f_1 of Example 15 is both injective and surjective. The function f_2 is neither injective nor surjective. The map f_3 is injective, but not surjective.

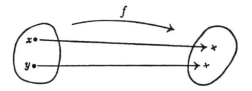

Figure 1.18. Two different points are always assigned to two different points.

Exercise. Find a function that is surjective but not injective.

Suppose that the map $f : \mathcal{M} \to \mathcal{N}$ is both injective and surjective. Then we can determine a unique $x \in \mathcal{M}$ that corresponds to $y \in \mathcal{N}$ by $f(x) = y$. (Why?) The map that assigns to each y this x is denoted by f^{-1}, and is called the *inverse map* of f. The inverse function of the map $f_1 : x \mapsto 2x - 1$ of Example 15 is $f_1^{-1} : x \mapsto \frac{1}{2}(x + 1)$.

Suppose that $\mathcal{M}, \mathcal{N},$ and $\mathcal{L}$ are three sets, that f is a map from $\mathcal{M}$ to $\mathcal{N}$, and that g is a map from $\mathcal{N}$ to $\mathcal{L}$. The map that assigns $g(f(x))$ to each element x of $\mathcal{M}$ is called the *composite map* of f and g, and is denoted by $g \circ f$:

$$g \circ f : \mathcal{M} \ni x \mapsto g(f(x)) \in \mathcal{L};$$

that is, $(g \circ f)(x) = g(f(x))$.

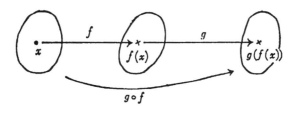

Figure 1.19.

Formula 1.2. Let $\mathcal{M}, \mathcal{N}, \mathcal{L},$ and $\mathcal{K}$ be four sets. Let $f, g,$ and h be maps from $\mathcal{M}$ to $\mathcal{N}$, $\mathcal{N}$ to $\mathcal{L}$, and $\mathcal{L}$ to $\mathcal{K}$, respectively. Then the following equality holds:

$$h \circ (g \circ f) = (h \circ g) \circ f$$

Proof: Both maps send x to $h(g(f(x)))$.   Q.E.D.

From now on we write $h \circ g \circ f$ instead of $h \circ (g \circ f)$ or $(h \circ g) \circ f$. If there are more maps involved, we can still write $f_5 \circ f_4 \circ f_3 \circ f_2 \circ f_1$ without parentheses ().

The map from $\mathcal{M}$ to itself assigning to each $x \in \mathcal{M}$ the element x itself, $x \mapsto x$, is called the *identity map* of $\mathcal{M}$. This is denoted by $id_\mathcal{M}$; i.e., $id_\mathcal{M}(x) = x$.

Formula 1.3. For any $f : \mathcal{M} \to \mathcal{N}$, $f \circ id_{\mathcal{M}} = f = id_{\mathcal{N}} \circ f$.

Formula 1.4. Let $f : \mathcal{M} \to \mathcal{N}$ and $g : \mathcal{N} \to \mathcal{L}$ be both surjective and injective. Then $g \circ f : \mathcal{M} \to \mathcal{L}$ is also both surjective and injective, and

$$(f^{-1})^{-1} = f,$$

$$f^{-1} \circ f = id_{\mathcal{M}}, \qquad f \circ f^{-1} = id_{\mathcal{N}},$$
$$(g \circ f)^{-1} = f^{-1} \circ g^{-1}$$

The proof is left to the reader.

The Second Week:
Equivalence Classes

(*Instructor's Soliloquy:* I do not like teaching abstract concepts at the beginning of a course, but since I have to make a choice, this is better than not being able to explain important concepts later.)

When we divide a set $\mathcal{M}$ into the sum of a number—either finite or infinite—of mutually disjoint subsets $\mathcal{N}_\alpha, \mathcal{N}_\beta, \mathcal{N}_\gamma, \ldots$, we say that we have a *partition* of $\mathcal{M}$ (see the figure).

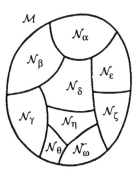

Figure 2.1.

You can easily check that the subsets $\mathcal{N}_\alpha, \mathcal{N}_\beta, \mathcal{N}_\gamma, \ldots$ satisfy the following conditions:

C1: $\mathcal{M} = \mathcal{N}_\alpha \cup \mathcal{N}_\beta \cup \mathcal{N}_\gamma \cup \cdots = \bigcup_\xi \mathcal{N}_\xi$ (ξ runs through the elements of the set $\{\alpha, \beta, \gamma, \ldots\}$). This means that the set $\mathcal{M}$ is completely covered by the subsets $\mathcal{N}_\alpha, \mathcal{N}_\beta, \mathcal{N}_\gamma, \ldots$.

C2: If $\mathcal{N}_\xi \neq \mathcal{N}_\eta$, then $\mathcal{N}_\xi \cap \mathcal{N}_\eta = \emptyset$. Here ξ and η are an arbitrary pair of elements from $\{\alpha, \beta, \gamma, \ldots\}$.

In short, a partition of a set $\mathcal{M}$ is a description of $\mathcal{M}$ as the union of those subsets $\mathcal{N}_\alpha, \mathcal{N}_\beta, \mathcal{N}_\gamma, \ldots$ satisfying C2. When a partition of $\mathcal{M}$ is given, each of the subsets $\mathcal{N}_\alpha, \mathcal{N}_\beta, \mathcal{N}_\gamma$ in the sum $\mathcal{M} = \bigcup_\xi \mathcal{N}_\xi$, is called a *class* by mathematicians. For nonmathematicians, a "compartment", a "country", or a "territory" might be easier to understand. We will use any of these words in this lecture.

Example 1. Suppose $\mathcal{M}$ is the set of all human beings on earth at this moment (= 1:36:27 p.m. on December 19, 1960). Let $\mathcal{N}_1$ be the set of males and $\mathcal{N}_2$ be the set of females. Then $\mathcal{M} = \mathcal{N}_1 \cup \mathcal{N}_2$ and $\mathcal{N}_1 \cap \mathcal{N}_2 = \emptyset$. Therefore, $\mathcal{M} = \mathcal{N}_1 \cup \mathcal{N}_2$ gives a partition of $\mathcal{M}$. The classes in this partition are $\mathcal{N}_1$ and $\mathcal{N}_2$.

Example 2. The set $\mathcal{M}$ is as in Example 1. Let $\mathcal{N}_0$ be the set of all babies of age 0, $\mathcal{N}_1$ be the set of children of age 1, $\mathcal{N}_2$ the set of children of age 2, ..., and, in general, $\mathcal{N}_k$ is the set of all people of age k. Clearly, $\mathcal{M} = \bigcup_{k=0}^{\infty} \mathcal{N}_k$, and $\mathcal{N}_k \cap \mathcal{N}_j = \emptyset$ if $\mathcal{N}_k \neq \mathcal{N}_j$. Therefore, $\mathcal{M} = \bigcup_{k=0}^{\infty} \mathcal{N}_k$ gives a partition.

It is convenient to use equivalence relations in order to find a partition of a set. Now let me explain what an equivalence relation is. First, a *binary relation* is a relation between two objects (or elements of a set). For example, the relation of being brothers, as in "Mr. X and Mr. Y are brothers", is such a relation.

Example 3. Some binary relations in the set of human beings.

(i) x and y are brothers. Let us write xBy to abbreviate this expression. For example, John Kennedy B Ted Kennedy, Groucho Marx B Harpo Marx, Tweedledee B Tweedledum, etc. When Mr. U and Mr. V are not brothers, we write $U\not\!\!B V$. For example, unfortunately Kuga $\not\!\!B$ Rockefeller.

(ii) We write xFy to mean x is the father of a boy y. The following theorem holds.
   Theorem. If xFy and xFz and $y \neq z$, then yBz.

(iii) x is in love with y. Notation: xLy. One of the misfortunes of life is that xLy does not imply that yLx. When there is a relation among three elements such as xLy and zLy, such a relation is called a *ternary relation*. Non-mathematicians would call this a love triangle.

Example 4. Let $\mathcal{L}$ be the set of all straight lines in three-dimensional Euclidean space E^3. Here are some binary relations between two elements of $\mathcal{L}$:

(i) l and l' are parallel. Notation: $l\|l'$; negation is $l \not\!\| l'$.

(ii) l and l' are perpendicular. Notation: $l \perp l'$; negation is $l \not\perp l'$.

(iii) l and l' intersect.

Here we define l and l' to be parallel if $l = l'$, or l and l' are contained in the same plane without intersecting. The following theorem is well known.

Theorem 2.1.

(1) $l\|l$

(2) $l\|l'$ implies $l'\|l$

(3) $l\|l'$ and $l'\|l''$ imply $l\|l''$.

---

(1) and (2) are obvious from the definition. (3) is not so direct, but the proof is easy.

If a binary relation $\sim$ on a set $\mathcal{M}$ satisfies the following three conditions E1,

E2, and E3, we call it an *equivalence relation*.

E1: For any element $x \in \mathcal{M}$, $x \sim x$.

E2: $x \sim y$ implies $y \sim x$

E3: $x \sim y$ and $y \sim z$ imply $x \sim z$.

Example 5. Some equivalence relations in the set of all humans.

(i) Have the same age.

(ii) Live in the same country.

(iii) Are of the same sex.

Example 6. In the previous example, if we stretch the meaning of "brothers" so that a person is his own brother, the relation B becomes an equivalence relation in the set of all males. E1: xBx (now that B has a new meaning), E2: xBy implies yBx, and E3: xBy and yBz imply xBz.

Example 7. In the set of straight lines $\mathcal{L}$, the relation $\parallel$ is an equivalence relation.

Now suppose $\sim$ is an equivalence relation on a set $\mathcal{M}$. We will denote by $\mathcal{N}(x)$ the subset of $\mathcal{M}$ consisting of all those elements y that satisfy $y \sim x$. Namely,

(1) $\mathcal{N}(x) = \{y \in \mathcal{M} \mid y \sim x\}$.

Of course,

(2) $x \in \mathcal{N}(x)$,

because $x \sim x$ (E1).

Thus when you choose x, the subset $\mathcal{N}(x)$ is determined. As you change x within $\mathcal{M}$, various $\mathcal{N}(x)$ emerge. But the union of all these subsets is, of course, $\mathcal{M}$. That is,

C1: $\mathcal{M} = \bigcup_{x \in \mathcal{M}} \mathcal{N}(x)$.

This is because any element y of $\mathcal{M}$ is contained in $\mathcal{N}(y)$.

C2: $\mathcal{N}(x) \neq \mathcal{N}(y)$ implies $\mathcal{N}(x) \cap \mathcal{N}(y) = \emptyset$.

Indeed, if $\mathcal{N}(x) \cap \mathcal{N}(y) \neq \emptyset$, then there is an element $z \in \mathcal{N}(x) \cap \mathcal{N}(y)$. By E2 and E3, $x \sim y$. Therefore, by E2 and E3 again, if $u \sim x$, then $u \sim y$ and vice versa. This means $\mathcal{N}(x) = \mathcal{N}(y)$. This proves C2.

What we have proved now is this: $\mathcal{M} = \bigcup_{x \in \mathcal{M}} \mathcal{N}(x)$ gives a partition of $\mathcal{M}$. This shows how an equivalence relation $\sim$ on $\mathcal{M}$ gives rise to a partition of $\mathcal{M}$. As we change x in $\mathcal{M}$, we obtain an $\mathcal{N}(x)$ corresponding to each x. However, the sets $\mathcal{N}(x)$ are not necessarily distinct. In other words, even if $x \neq y$ it is possible that $\mathcal{N}(x) = \mathcal{N}(y)$, as we have seen. For simplicity, assume $\mathcal{M}$ is finite. Although it seems that the number of sets $\mathcal{N}(x)$ is the same as the number of elements of $\mathcal{M}$, denoted $|\mathcal{M}|$, there may be much repetition of the sets $\mathcal{N}(x)$. So, the number of distinct $\mathcal{N}(x)$'s is usually considerably smaller than $|\mathcal{M}|$. In the expression $\mathcal{M} = \bigcup_{x \in \mathcal{M}} \mathcal{N}(x)$ there is much waste, though the statement is certainly correct. See the following example.

Example 8. Let $\mathcal{M}$ be the set of all humans, and $x \sim y$ be the equivalence relation "x and y are of the same sex". Then it follows

Kuga $\sim$ Albert Einstein

Richard Nixon $\sim$ Elvis Presley

Yoko Ono $\sim$ Shirley MacLaine

Ann Landers $\sim$ Abigail van Buren $\not\sim$ Kuga.

If we make different $\mathcal{N}(x)$ by changing x, we have

$$\mathcal{N}(\text{Kuga}) = \mathcal{N}(\text{Einstein}) = \mathcal{N}(\text{Nixon}) = \mathcal{N}(\text{Mao Tse Tung}) = \mathcal{N}(\text{Elvis}) = \dots.$$

We also have

$$\mathcal{N}(\text{Kyoko Kuga}) = \mathcal{N}(\text{Shirley MacLaine}) = \mathcal{N}(\text{Brigitte Bardot}) = \dots.$$

Each of these is identical to either $\{\text{men}\} = \mathcal{N}_1$ or $\{\text{women}\} = \mathcal{N}_2$, and there are only two distinct classes. In order to simplify the expression of $\mathcal{M}$ as the union of $\mathcal{N}(x)$, simply choose one representative of all men and one representative of all women; for example, $\mathcal{M} = \mathcal{N}(\text{Michio Kuga}) \cup \mathcal{N}(\text{Shirley MacLaine})$. In general, let us consider the set of all different compartments in a partition $\mathcal{M} = \bigcup_x \mathcal{N}(x)$ which is generated by an equivalence relation $\sim$. This set is denoted by $\mathcal{M}/\sim$ and is called the *quotient space* (or *quotient set*) of $\mathcal{M}$ by $\sim$. (An element of $\mathcal{M}/\sim$ is itself a subset $\mathcal{N}(x)$. The set $\mathcal{M}/\sim$ is a set of sets.)

Example 9. Let $\mathcal{M}$ be the set of all humans, and $\sim$ the equivalence relation of being of the same sex. Then $\mathcal{M}/\sim$ consists of only two elements $\mathcal{N}_1$ and $\mathcal{N}_2$, i.e., $\mathcal{M}/\sim = \{\mathcal{N}_1, \mathcal{N}_2\}$.

Example 10. Let $\mathcal{L}$ be the set of straight lines as before. We want to find a partition of $\mathcal{L}$ by the equivalence relation $\|$ (parallelness). Each equivalence class consists of parallel lines. We call each class a "direction". (Actually, an equivalence class is a pair consisting of a direction and its opposite, since a line that points northwest also points southeast. So, $\mathcal{L}/\|$ is the set of all directions.

The map which assigns the class $\mathcal{N}(x)$ to each element x of $\mathcal{M}$ is a map from $\mathcal{M}$ to $\mathcal{M}/\sim$. This is obviously a surjection. We call this map the *natural map*. If we denote this map by ν,

$$\mathcal{M} \ni x \iff \nu(x) = \mathcal{N}(x).$$

(Soliloquy: This is not thorough enough, but maybe there will be a second chance. I wonder if the students understood the subtlety involved in putting the statement of C2 as $\mathcal{N}_\alpha \neq \mathcal{N}_\beta$ implies $\mathcal{N}_\alpha \cap \mathcal{N}_\beta = \emptyset$ instead of using $\alpha \neq \beta$ implies $\mathcal{N}_\alpha \cap \mathcal{N}_\beta = \emptyset$?!)

Exercise. Let f be a map from $\mathcal{M}$ to $\mathcal{N}$. If we define $\sim$ by $x \sim y$ if and only if $f(x) = f(y)$, then $\sim$ is an equivalence relation on $\mathcal{M}$. Show that every equivalence relation coincides with some equivalence relation $\sim$ defined in this way using some function f.

Answer: $\mathcal{N} = \mathcal{M}/\sim$, $f = \nu$ (natural map).

The Third Week:
The Story of Free Groups

We shall start with a set of $2n + 1$ "letters",

$$E; A_1, A_2, A_3, A_4, ..., A_n,$$
$$A_1^{-1}, A_2^{-1}, A_3^{-1}, A_4^{-1}, ..., A_n^{-1}.$$

An arbitrary sequence of these letters is called a *word*. For example,

$$A_5 A_3 A_1^{-1} A_{100} A_{29} E A_{29}^{-1} E A_{91} A_{91} \text{ (a word of length 10)},$$

$$A_4 A_2^{-1} A_2 A_2 A_2 A_1^{-1} A_1^{-1} E E E A_{101} \text{ (length 11)},$$

$$A_1 A_2^{-1} \text{ (length 2)},$$

$$A_1^{-1} \text{ (length 1)},$$

$$E \text{ (length 1)},$$

are all words. The number of letters used in a word is the length of the word. If A_i occurs m times, we count it m times. We denote the set of all words by $\mathcal{W}$. This is an infinite set.

Warning: Needless to say, two words containing exactly the same set of letters can be different if the orders of these letters are different. For example,

$$A_1 A_3^{-1} A_5 E A_2 A_6 E A_5 A_2$$
$$\text{and}$$
$$A_5 A_5 A_2 A_6 E A_3^{-1} E A_1 A_2$$

are two different words. We use $W_1, W_2, ...$ to denote elements (*i.e.*, words) of $\mathcal{W}$. A word is itself a sequence of letters, but we consider this sequence as a single object and represent it by a single letter such as W. This is very common in mathematics. For example, a complex number $x + yi$ is determined only when you specify two real numbers x and y; we often express such a number by one letter z. Also, we use $\mathbf{x}$ or $\vec{x}$ to represent a vector $(x_1, x_2, ..., x_n)$.

Given two words such as $A_1 A_3^{-1} A_5 E A_4 A_2^{-1}$ and $A_3 A_1 A_7^{-1} E A_2$, we can join them to make a new and longer word,

$$A_1 A_3^{-1} A_5 E A_4 A_2^{-1} A_3 A_1 A_7^{-1} E A_2.$$

We call this longer word the *product* (or *concatenation*) of $A_1 A_3^{-1} A_5 E A_4 A_2^{-1}$ and $A_3 A_1 A_7^{-1} E A_2$. In general, the product of two words W_1 and W_2 is the word obtained by joining W_1 and W_2. We use the symbol $W_1 \cdot W_2$ to denote the product of W_1 and W_2.

Caution: $W_1 \cdot W_2$ and $W_2 \cdot W_1$ are not necessarily the same word. For example, if $W_1 = A_1 A_3^{-1} A_5 E A_4 A_2^{-1}$ and $W_2 = A_3 A_1 A_7^{-1} E A_2$,

$$W_1 \cdot W_2 = A_1 A_3^{-1} A_5 E A_4 A_2^{-1} A_3 A_1 A_7^{-1} E A_2,$$

but

$$W_2 \cdot W_1 = A_3 A_1 A_7^{-1} E A_2 A_1 A_3^{-1} A_5 E A_4 A_2^{-1}.$$

As you see, $W_1 \cdot W_2 \neq W_2 \cdot W_1$.

It is obvious that

Formula 3.1. $(W_1 \cdot W_2) \cdot W_3 = W_1 \cdot (W_2 \cdot W_3)$

---

holds. From now on we denote the product of three words by $W_1 \cdot W_2 \cdot W_3$ without parentheses. Of course, this is nothing but the three words W_1, W_2, and W_3 joined. Similarly, the product $W_1 \cdot W_2 \cdot W_3 \cdot \cdots \cdot W_m$ is the expression obtained by joining these words.

Next we want to define some operations on elements of $\mathcal{W}$.

(I) If the expression $A_i A_i^{-1}$ occurs within a word, erase these two letters and replace them with E. This is an operation of type I.

(operation of type I)

(II) Similarly if $A_i^{-1} A_i$ occurs in a word, replace these by E.

(operation of type II)

(III) The reverse of an operation of type I. That is, replace the letter E by $A_i A_i^{-1}$. The index i can be anything between one and n.

(operation of type III)

(IV) Similarly, the reverse of an operation of type II is an operation of type IV. That is, replace E by $A_i^{-1}A_i$.

$$\underbrace{* * \cdots *}\ E\ \underbrace{* * \cdots *}$$
$$\| \quad \Downarrow \quad \|$$
$$\overbrace{* * \cdots *}\,A_i^{-1}A_i\,\overbrace{* * \cdots *}$$
                                        (operation of type IV)

(V) If a word has length ≥ 2, and if it contains E, erase E.

$$\underbrace{* * \cdots *}E\underbrace{* * \cdots *}$$
$$\| \quad \Downarrow \quad \|$$
$$\overbrace{* * \cdots *}\ \overbrace{* * \cdots *}$$
                                        (operation of type V)

Caution: Of course, after E is erased, there will be a space of length 1, so we must compress the word to eliminate this space.

(VI) The reverse of the preceding operation. That is, insert the letter E between any two letters of a word (or at the beginning or the end of the word). For example,

$$EA_6A_1A_3^{-1}A_5EA_8^{-1} \Leftarrow A_6A_1A_3^{-1}A_5EA_8^{-1} \Rightarrow A_6A_1A_3^{-1}A_5EA_8^{-1}E$$
$$\Downarrow$$
$$A_6A_1A_3^{-1}EA_5EA_8^{-1}$$
$$\Downarrow$$
$$A_6A_1A_3^{-1}A_5EEA_8^{-1}$$

We call a combination (composition) of a finite number of these six operations a *fundamental transformation*. When the word W_1 can be changed to W_2 by a fundamental transformation, we say that W_1 and W_2 are *equivalent*, and denote this by $W_1 \sim W_2$. For example, the following fundamental transformations

$$A_{10}A_3^{-1}EA_3A_2^{-1} \Rightarrow A_{10}A_3^{-1}A_3A_2^{-1} \Rightarrow A_{10}EA_2^{-1} \Rightarrow A_{10}A_2^{-1}$$
$$\Downarrow$$
$$A_{10}A_3^{-1}A_1^{-1}A_1A_3A_2^{-1}$$
$$\Downarrow$$
$$A_{10}EA_3^{-1}A_1^{-1}A_1A_3A_2^{-1}$$
$$\Downarrow$$
$$A_{10}A_5A_5^{-1}A_3^{-1}A_1^{-1}A_1A_3A_2^{-1}$$
$$\Downarrow$$
$$EA_{10}A_5A_5^{-1}A_3^{-1}A_1^{-1}A_1A_3A_2^{-1}$$

show that

$$A_{10}A_3^{-1}EA_3A_2^{-1} \sim A_{10}A_2^{-1},$$

and
$$A_{10}A_3^{-1}EA_3A_2^{-1} \sim EA_{10}A_5A_5^{-1}A_3^{-1}A_1^{-1}A_1A_3A_2^{-1}.$$

Obviously we have

Proposition 3.1. The relation $\sim$ above is an equivalence relation. That is,

(a) $W_1 \sim W_1$

(b) $W_1 \sim W_2 \Rightarrow W_2 \sim W_1$

(c) $W_1 \sim W_2$ and $W_2 \sim W_3 \Rightarrow W_1 \sim W_3$

We also have

Proposition 3.2.

(d) $\qquad \left. \begin{array}{l} W_1 \sim W_2 \\ W_3 \sim W_4 \end{array} \right\} \Rightarrow W_1 \cdot W_3 \sim W_2 \cdot W_4$

You can prove these very easily.

Example. Let
$$W_1 = A_{10}A_3^{-1}EA_3A_2^{-1}, \quad W_2 = A_{10}A_2^{-1}$$
$$W_3 = A_2A_5EA_5^{-1}A_6^{-1}, \quad W_4 = A_2A_6^{-1}.$$
Then $W_1 \sim W_2$ and $W_3 \sim W_4$.

Now
$$W_1 \cdot W_3 = A_{10}A_3^{-1}EA_3A_2^{-1}A_2A_5EA_5^{-1}A_6^{-1} \sim A_{10}A_2^{-1}A_2A_6^{-1}$$
$$= W_2 \cdot W_4.$$

Next is a procedure for producing a new word from a word W. We replace each letter A_i in W by A_i^{-1}, A_i^{-1} by A_i, and keep E untouched. Then we reverse the order of the letters in the word. The new word we obtain is denoted by W^{-1}. For example, if
$$W = A_1A_5^{-1}A_3A_4^{-1}A_2EA_5A_3^{-1}E,$$
then
$$W^{-1} = EA_3A_5^{-1}EA_2^{-1}A_4A_3^{-1}A_5A_1^{-1}.$$

Obviously,

Proposition 3.3.
$$(W^{-1})^{-1} = W$$

Proposition 3.4.
$$W_1 \sim W_2 \implies W_1^{-1} \sim W_2^{-1}$$

Proposition 3.5.
$$W \cdot W^{-1} \sim W^{-1} \cdot W \sim E.$$

Here, E is the word consisting of the letter E alone.

For example, when
$$W = A_1 A_5^{-1} A_3 A_4^{-1} A_2 E A_5 A_3^{-1} E,$$
then
$$W^{-1} = E A_3 A_5^{-1} E A_2^{-1} A_4 A_3^{-1} A_5 A_1^{-1}.$$
Therefore,
$$
\begin{aligned}
W \cdot W^{-1} &= A_1 A_5^{-1} A_3 A_4^{-1} A_2 E A_5 A_3^{-1} E E A_3 A_5^{-1} E A_2^{-1} A_4 A_3^{-1} A_5 A_1^{-1} \\
&\sim A_1 A_5^{-1} A_3 A_4^{-1} A_2 E A_5 A_3^{-1} A_3 A_5^{-1} E A_2^{-1} A_4 A_3^{-1} A_5 A_1^{-1} \\
&\sim A_1 A_5^{-1} A_3 A_4^{-1} A_2 E A_5 E A_5^{-1} E A_2^{-1} A_4 A_3^{-1} A_5 A_1^{-1} \\
&\sim A_1 A_5^{-1} A_3 A_4^{-1} A_2 E A_5 A_5^{-1} E A_2^{-1} A_4 A_3^{-1} A_5 A_1^{-1} \\
&\sim A_1 A_5^{-1} A_3 A_4^{-1} A_2 E E E A_2^{-1} A_4 A_3^{-1} A_5 A_1^{-1} \\
&\sim A_1 A_5^{-1} A_3 A_4^{-1} A_2 A_2^{-1} A_4 A_3^{-1} A_5 A_1^{-1} \\
&\sim A_1 A_5^{-1} A_3 A_4^{-1} A_4 A_3^{-1} A_5 A_1^{-1} \\
&\sim A_1 A_5^{-1} A_3 A_3^{-1} A_5 A_1^{-1} \\
&\sim A_1 A_5^{-1} A_5 A_1^{-1} \\
&\sim A_1 A_1^{-1} \\
&\sim E.
\end{aligned}
$$

Similarly, $W^{-1} \cdot W = E$. The general situation is the same. Q.E.D.

Let us denote the quotient set $\mathcal{W}/\sim$ of $\mathcal{W}$ by this equivalence relation $\sim$ by the symbol F; i.e., $F = \mathcal{W}/\sim$. We denote elements of F by $w_1, w_2, \ldots, w_p, \ldots$. Each of these is an equivalence class by $\sim$.

Suppose w_1 and w_2 are two elements of F. Choose any word W_1 from w_1 and W_2 from w_2. Make the product $W_1 \cdot W_2$, and denote the equivalence class to which $W_1 \cdot W_2$ belongs by w. The procedure is summarized as follows:

$$
\left.
\begin{aligned}
w_1 &\ni W_1 \text{ (arbitrary choice)} \\
w_2 &\ni W_2 \text{ (arbitrary choice)}
\end{aligned}
\right\} \longrightarrow \text{product } W_1 \cdot W_2
$$

$$\longrightarrow \text{equivalence class of } W_1 \cdot W_2 = w.$$

Since both w_1 and w_2 contain an infinite number of elements (words), let's change W_1 to another W_1' still belonging to w_1, and replace W_2 by W_2' from w_2.

$$\left.\begin{array}{c} w_1 \ni W_1' \\ w_2 \ni W_2' \end{array}\right\} \longrightarrow W_1' \cdot W_2' \longrightarrow \text{equivalence class of } W_1' \cdot W_2' = w'.$$

Actually, the equivalence class w' is nothing but w. This is because W_1 and W_1' belong to the same equivalence class and W_2 and W_2' belong to the same equivalence class, so $W_1 \sim W_1'$ and $W_2 \sim W_2'$. Proposition 3.2 states that $W_1 \cdot W_2 \sim W_1' \cdot W_2'$. Therefore, both products belong to the same equivalence class, namely $w = w'$. Q.E.D.

This means that no matter which representatives W_1 and W_2 you choose from w_1 and w_2 respectively, the equivalence class to which $W_1 \cdot W_2$ belongs is w. That is, the class w is determined by w_1 and w_2, and does not depend on the choice of W_1 and W_2. We can now call w the product of w_1 and w_2, and adopt the notation

$$w_1 \cdot w_2.$$

Now we have defined a product

$$\left.\begin{array}{c} F \ni w_1 \\ F \ni w_2 \end{array}\right\} \longrightarrow w_1 \cdot w_2 \in F$$

in the set F. The class which contains the word consisting of the letter E alone is denoted by e. Since

$$A_1 A_1^{-1} \sim E,$$

then

$$E \in e, \ A_i A_i^{-1} \in e, \ A_i^{-1} A_i \in e, \ EE \in e, \text{ etc.}$$

Similarly, we choose a representative W from a class w, and make the word W^{-1}. We denote by w^{-1} the class to which W^{-1} belongs. As before, w^{-1} does not depend on the choice of W; it depends on w alone (by Proposition 3.4).

Theorem 3.1. If w_1, w_2, w_3, $w \in F$, we have

    (1) $(w_1 \cdot w_2) \cdot w_3 = w_1 \cdot (w_2 \cdot w_3)$

    (2) $w \cdot e = e \cdot w = w$

    (3) $w \cdot w^{-1} = w^{-1} \cdot w = e$

The proof is left to the reader.

To those who know the definition of groups, it is better to say

> **Theorem 3.2.** The set F is a group under product. The unit element is e. The inverse of w is w^{-1}.

In the 0th week, I recommended that you teach yourself the concept of groups by this week. Have you done it? We will now use some terms from group theory.

As I said, the set F becomes a group. This group is called the *free group generated by* A_1, $A_2, \ldots,$ and A_n. The letters A_1, $A_2, \ldots,$ A_n are called the *generators* of the group. When you want to emphasize the fact that F is generated by A_1, $A_2, \ldots,$ A_n, F is written as $F(A_1,\ A_2, \ldots,\ A_n)$. The element of F containing the one-letter word A_i is denoted by the same symbol A_i.

Finally, an interesting theorem. (Prove it yourself.)

> **Theorem 3.3.** Let G be an arbitrary group, and let a_1, $a_2, \ldots,$ a_n be any elements of G. Then there is a unique homomorphism φ from $F(A_1,\ A_2, \ldots,\ A_n)$ to G satisfying
> $$\varphi(A_1) = a_1,\ \varphi(A_2) = a_2, \ldots, \varphi(A_n) = a_n$$

Hint: Let $\varphi(A_3 A_5^{-1} A_6 E A_7^{-1}) = a_3 a_5^{-1} a_6 a_7^{-1}$, for example.

Heave Ho! (Pull it tight)

The Fourth Week:
The fundamental group of a surface

The scene of today's lecture is set in a region in a plane. We define a region to be a part of a plane surrounded by some closed curves. For example, the portion D in Figure 4.1 surrounded by the closed curves C_1, C_2, and C_3 is a region (i.e., the unshaded part of the figure).

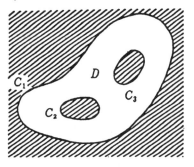

Figure 4.1.

In the lectures we often call a region "the land" and the outside of a region "the sea", or "lakes", or "ponds". The main character in today's story is a curve. The curves in the lectures will be limited to those that have two end points (one initial point and one terminal point), and have a direction (or *orientation*) from the initial point to the terminal point. For example, the "curve" C in Figure 4.2, which approaches a circle infinitely closely, is excluded from the curves under consideration because C does not have a terminal point.

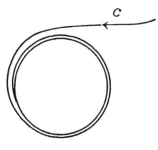

Figure 4.2.

However, a curve could be zigzag, straight, or even have intersection with itself.

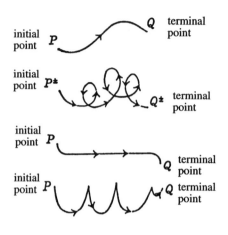

Figure 4.3.

(See Figure 4.3.)

It is not necessary that the initial point and the terminal point be different. When the two endpoints coincide, we call the curve a *closed curve*.

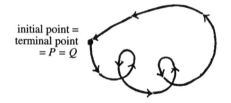

Figure 4.4.

Since we are considering oriented curves only, we must consider the curve obtained from a curve C by reversing the orientation as distinct from C itself. We express the curve with reversed orientation by C^{-1}.

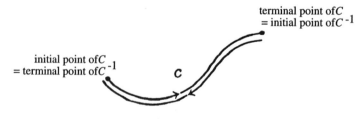

Figure 4.5.

Obviously,

$$\text{the initial point of } C = \text{ the terminal point of } C^{-1}$$

$$\text{the terminal point of } C = \text{ the initial point of } C^{-1}$$

When the terminal point of C_1 coincides with the initial point of C_2, we can obtain a third curve by joining C_1 and C_2. Of course, this is the curve obtained by tracing C_1 first, then tracing C_2. We call this new curve the *product* (or *concatenation*) of C_1 and C_2, and denote it by $C_1 \cdot C_2$.

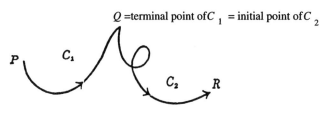

Figure 4.6.

It is clear that we have the following formula when the terminal point of C_1 is the initial point of C_2, and the terminal point of C_2 is the initial point of C_3.

Formula 4.1.
$$(C_1 \cdot C_2) \cdot C_3 = C_1 \cdot (C_2 \cdot C_3)$$

Both sides of Formula 4.1 represent a curve which traces C_1, C_2, and C_3 in this order. From now on we write $C_1 \cdot C_2 \cdot C_3$ to denote this curve.

If you can always draw a curve in D from P to Q for an arbitrary pair of points P and Q, we say that D is *connected*. In other words, it is not always possible to travel from P to Q without swimming, jumping, or flying. We will consider only connected regions in the rest of the lecture.

Figure 4.7.

Now we consider the set of curves in a connected region D. We denote the set of curves in D by $W(D)$. Then we define an equivalence relation $\sim$ as follows:

Two curves C_1 and C_2 (both elements of $W(D)$) are equivalent (i.e., $C_1 \sim C_2$) if:

H1: The initial point of C_1 is the initial point of C_2;

H2: The terminal point of C_1 is the terminal point of C_2;

H3: C_1 can be deformed continuously to C_2 without moving the end points.

In other words, we consider the curve C_1 to be made of a rubber string. Then move C_1 around D, while keeping the end points fixed, by stretching or contracting the rubber string C_1. Be careful to keep C_1 strictly inside D while you do this. Never let the string C_1 get wet in the lakes.

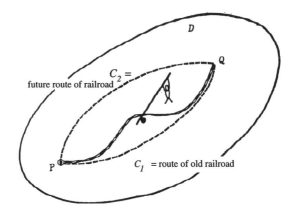

Figure 4.8. Men at work. The railroad is deforming continuously.

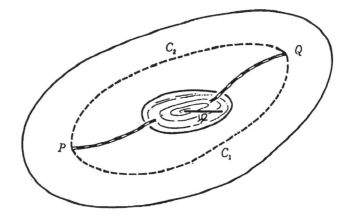

Figure 4.9. You can't deform continuously if there is a pond.

It is obvious that the relation $\sim$ is an equivalence relation. Therefore, we have an equivalence relation $\sim$ on $W(D)$. Mathematicians say that C_1 and C_2 are *homotopic* when $C_1 \sim C_2$. In Figure 4.10, C_1, C_2, and C_3 are all homotopic to each other, but C_1 and C_4 are not. We will define homotopy groups next week based on this idea.

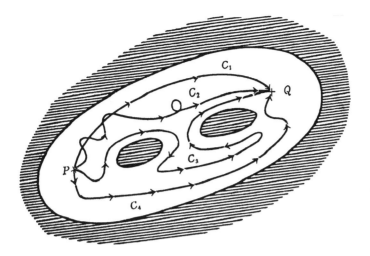

Figure 4.10.

Note: This discussion on curves is qualitative and pictorial. In order to work with curves, it will be necessary to view them as equivalence classes of continuous functions defined on an interval. Two equivalent curves look exactly the same, but may be parametrized by different functions.

The Fifth Week:
Fundamental Groups

Last week we were considering curves in a region. A region was a part of the plane surrounded by closed curves (for example, the unshaded part of figure 5.1). We decided to call such a region land and the outside region a sea or a lake (such as the shaded part of Figure 5.1) .

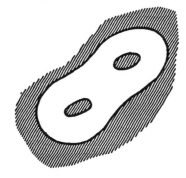

Figure 5.1.

When we considered a curve in the region D, we always meant a curve with an initial point, a terminal point, and orientation. In these lectures, a region means a connected region; namely, any two points in the region can be connected by a curve in the region.

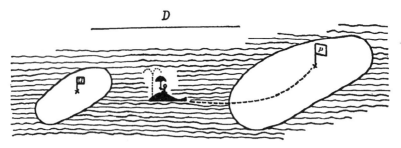
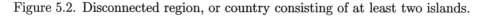

Figure 5.2. Disconnected region, or country consisting of at least two islands.

When C_1 and C_2 are two curves in D, we call them homotopic, and use the notation $C_1 \sim C_2$, if the following conditions are satisfied:

H1: The initial point of C_1 is the initial point of C_2.

H2: The terminal point of C_1 is the terminal point of C_2.

H3: C_1 can be deformed continuously into C_2 without moving the end points.

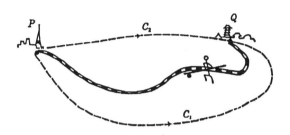

Figure 5.3. Deforming C_1 to C_2 continuously.

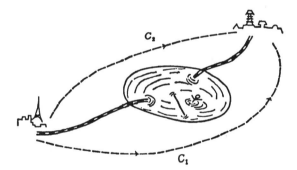

Figure 5.4. You can't deform continuously if there is a pond: $C_1 \not\sim C_2$.

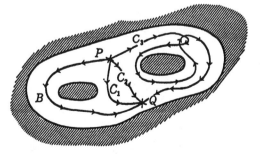

Figure 5.5. $C_1 \sim C_2 \sim C_3 \not\sim B$.

So far this is a review of the last week.

It is clear that we have the following proposition:

Proposition 5.1. If $C_1 \sim C_2$, $C_3 \sim C_4$, and the terminal point of C_1 the initial point of C_3, then $C_1 \cdot C_3 \sim C_2 \cdot C_4$.

Recall that $C_1 \cdot C_3$ is the product curve obtained by tracing C_1 first, then C_3. Try to prove this proposition by looking at Figure 5.6.

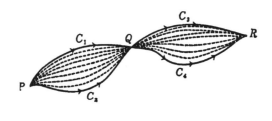

Figure 5.6. $\left.\begin{matrix} C_1 \sim C_2 \\ C_3 \sim C_4 \end{matrix}\right\} \implies C_1 \cdot C_3 \sim C_2 \cdot C_4.$

The following proposition is clear, too.

Proposition 5.2. If $C_1 \sim C_2$, then $C_2^{-1} \sim C_1^{-1}$.

The symbol C^{-1} stands for the curve C with the orientation reversed.

Now we consider a subset of $W(D)$. Choose a point O in D. Consider the set $W(D; O)$ of closed curves in D with O as both the initial and terminal point. (See Figure 5.7)

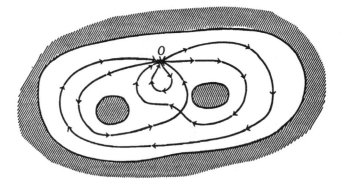

Figure 5.7. Closed curves which belong to $W(D; O)$

The product of the elements C_1 and C_2 of $W(D;O)$ is a curve which starts at O, returns to O once by way of C_1, and returns again to O via C_2. Therefore, $C_1 \cdot C_2$ is a closed curve starting (and ending) at O, so $C_1 \cdot C_2$ is again an element of $W(D;O)$.

Lemma 5.1. If C_1 and $C_2 \in W(D;O)$, then $C_1 \cdot C_2 \in W(D;O)$.

It is also clear that we have

Formula 5.1.
$$(C_1 \cdot C_2) \cdot C_3 = C_1 \cdot (C_2 \cdot C_3)$$

See Formula 4.1, which is similar.

Now let us consider the quotient space, $W(D;O)/\sim$, of the set $W(D;O)$ by the equivalence relation $\sim$ of homotopy. It is denoted by $\pi_1(D;O)$; namely, $\pi_1(D;O) = W(D;O)/\sim$. Each element of $\pi_1(D;O)$ is an equivalence class of closed curves from O to O in D. Such an equivalence class is called a *homotopy class*. The homotopy class which contains the closed curve C is denoted $[C]$. Let a and b be arbitrary elements of $\pi_1(D;O)$. By definition, each of a and b is a set of closed curves, all of which are homotopic to each other. Now choose closed curves A from a, and B from b. (Of course, we have $a = [A]$ and $b = [B]$.) Then, since the product $A \cdot B$ is again a closed curve starting and ending at O, (*i.e.*, $A \cdot B \in W(D;O)$), the curve $A \cdot B$ determines a homotopy class $[A \cdot B] \in \pi_1(D;O)$. Let $c = [A \cdot B]$. We can verify that this class c is completely determined as soon as the classes a and b are chosen, and it does not depend on which curves A and B we have chosen from a and b respectively. Indeed, choose A' from a and B' from b. Then the product $A' \cdot B' \sim A \cdot B$ by Proposition 5.1, because $A \sim A'$ and $B \sim B'$. That is, both $A \cdot B$ and $A' \cdot B'$ belong to the same homotopy class.

$$[A \cdot B] = [A' \cdot B'] = C$$

Therefore, we will use the symbol $a \cdot b$ to denote c, because c is determined by a and b alone, and call it the product of homotopy class a and b. In other words,

$$[A \cdot B] = a \cdot b.$$

This defines an operation which assigns to an arbitrary pair of homotopy classes a and b a third homotopy class $a \cdot b$. Let me use a diagram to summarize the situation:

$$\left. \begin{array}{l} \pi_1(D;O) \ni a \mapsto \text{ choose } A(\in a) \\ \pi_1(D;O) \ni b \mapsto \text{ choose } B(\in b) \end{array} \right\} \longrightarrow A \cdot B \longrightarrow [A \cdot B] = a \cdot b \in \pi_1(D;O)$$

The following is clear for a, b, and c in $\pi_1(D; O)$:

Proposition 5.3.

$$(a \cdot b) \cdot c = a \cdot (b \cdot c)$$

When a closed curve having O as its end points can be deformed continuously to the point O, we call the curve *null-homotopic*.

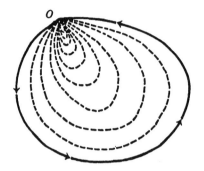

Figure 5.8. A null-homotopic curve can be contracted to O.

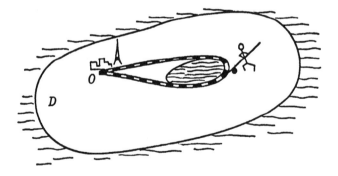

Figure 5.9.   A loop railroad which is not null-homotopic.

We denote the homotopy class consisting of all null-homotopic curves by 1. Obviously, the following is true for any $a \in \pi_1(D; O)$:

Proposition 5.4.

$$a \cdot 1 = 1 \cdot a = a$$

The inverse class of a homotopy class a is defined as $[A^{-1}]$ where A is a curve in a. We denote this by $a^{-1} : a^{-1} = [A^{-1}]$. The class a^{-1} depends only on the homotopy class a and is independent of the choice of A. The proof is obvious from Proposition 5.2.

For any $a \in \pi_1(D; O)$,

Proposition 5.5. For any $a \in \pi_1(D; O)$,

$$a \cdot a^{-1} = a^{-1} \cdot a = 1$$

Proof: Take $a \in A$. Since $a^{-1} = [A^{-1}]$, $a \cdot a^{-1} = [A \cdot A^{-1}]$. $A \cdot A^{-1}$ is a closed curve starting from O, stopping at O on the way, and finally coming back to O, following A in reverse. (See Figure 5.10.) Let's deform this continuously. Since all we need to fix are two end points, we can move the points in between without any restriction. The curve B, which follows A just until the point O comes into view, then returns, is equivalent to $A \cdot A^{-1}$, as in Figure 5.11. Similarly the curves B', B'', ... of Figures 5.12, 5.13, and 5.14 are all homotopic, and

$$A \cdot A^{-1} \sim B \sim B' \sim B'' \sim \ldots.$$

Finally, the curve $A \cdot A^{-1}$ will be deformed continuously to the point O. Therefore, $A \cdot A^{-1} = 1$ (= the constant curve) and $a \cdot a^{-1} = 1$. The argument is similar for $a^{-1} \cdot a$.   Q.E.D.

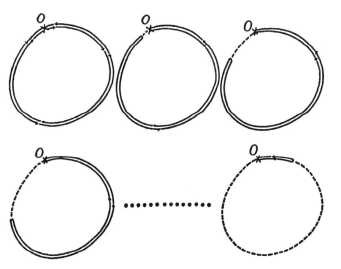

Figures 5.10, 5.11, 5.12, 5.13, 5.14.

Using the language of groups,

Theorem 5.1. The set of homotopy classes $\pi_1(D; O)$ is a group under product $a, b \mapsto a \cdot b$.

The unit is 1. The group $\pi_1(D; O)$ is called the *fundamental group* of D relative to the base point O. (Sometimes it is called the Poincaré group or the first homotopy group.) When we change the point O to another point O', what happens? There is a theorem.

Theorem 5.2. Let D be a connected region, and O, O' two points of D. Then $\pi_1(D; O)$ and $\pi_1(D; O')$ are isomorphic.

The proof is left to the reader. The idea is that for a curve A connecting O and O', the assignment of $[A^{-1} \cdot C \cdot A] \in \pi_1(D; O')$ to $[C]$ in $\pi_1(D; O)$ gives an isomorphism. (See Figure 5.15.)

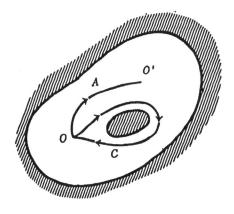

Figure 5.15.

When only the abstract group structure is considered, $\pi_1(D; O)$ is denoted by $\pi_1(D)$. When $\pi_1(D) = \{1\} =$ the trivial group, we say that D is *simply connected*. This means that any closed curve in D can be contracted to a point.

The Sixth Week:
Examples of fundamental groups

Example 1.   The region D is the plane minus one point. There is no sea, but there is an infinitely small lake P_0. (See Figure 6.1.) According to our rule, we are not allowed to move the curve across P_0. No matter how small, a lake is a lake, so we may not move our rubber railroad across P_0.

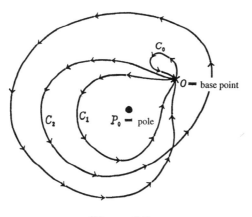

Figure 6.1.

If this is difficult to visualize, think of P_0 instead as an infinitely tall pole reaching to heaven. (See Figure 6.2).

Let C_1 be the curve at O and going around the pole once counterclockwise before coming back to O. Let C_2 be a similar closed curve that goes around the pole twice. In general, C_n is a closed curve that goes around the pole n times ($n = 1, 2, \ldots$). Also, we define C_{-n} as a closed curve going around the pole n times clockwise. Finally, let C_0 be a closed curve that is null-homotopic. Intuitively it is clear that C_1 is not homotopic to C_0, ($C_1 \not\sim C_0$), C_2 is not homotopic to C_1, ($C_2 \not\sim C_1$), nor is C_2 homotopic to C_0, ($C_2 \not\sim C_0$). If you are not convinced, tie a rope around a pole twice, and heave ho! Pull it tight! If the rope comes loose, then $C_2 \sim C_0$. But this won't happen.

In general, if $n \neq m$, then $C_n \not\sim C_m$. So if we define $c_n = [C_n]$,

$$c_0 = 1, \ c_1, c_2, \ldots, c_n, \ldots ; c_{-1}, c_{-2}, c_{-3}, \ldots$$

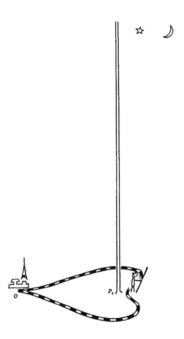

Figure 6.2.

are distinct elements of $\pi_1(D;O)$. It can be proved that there are no other elements in $\pi_1(D;O)$ than c_n $(n = 0, \pm1, \pm2, \ldots)$:

$$\pi_1(D;O) = \{c_0,\ c_1, c_2, \ldots; c_{-1}, c_{-2}, \ldots\}$$

Intuitively this is obvious, but a rigorous proof is difficult (and will not be given here). It is also clear that $C_m \cdot C_n = C_{m+n}$. Therefore, $\pi_1(D;O)$ is isomorphic to $\mathbf{Z}$ (the additive group of integers): $\pi_1(D;O) \simeq \mathbf{Z}$. In particular, it is an abelian group.

In order to write down the correspondence that gives this isomorphism, we do the following: Let $\mathbf{C}$ be the complex plane, and let a be the complex number corresponding to P_0. Since our closed curve C does not pass through the point P_0, the complex line integral $\displaystyle\int_C \frac{dz}{z-a}$ makes sense. (We are assuming that C is rectifiable, *i.e.*, C has length). If we let

$$n(C) = \frac{1}{2\pi i} \int_C \frac{dz}{z-a},$$

then $n(C)$ is an integer, and $C \sim C'$ implies that $n(C) = n(C')$. Therefore, the correspondence

$$W(D;O) \ni C \mapsto n(C) = \frac{1}{2\pi i} \int_C \frac{dz}{z-a} \in \mathbf{Z}$$

gives an isomorphism. (Look this up in any book on the theory of complex functions.)

This example shows the usefulness of fundamental groups in analysis.

Example 2. $D =$ the plane $- \{P_0, Q_0\}$, the plane with two points removed. Is the fundamental group of D commutative?

Answer: No, it is not. In order to see this, let P be a closed curve going around a pole P_0 once counterclockwise, and let Q be a closed curve around a pole Q_0 counterclockwise. Now it is sufficient to show that $P \cdot Q \not\sim Q \cdot P$. This is equivalent to showing $P \cdot Q \cdot P^{-1} \cdot Q^{-1} \not\sim 1$.

To see this, wind a rope around the two poles so that the rope represents $P \cdot Q \cdot P^{-1} \cdot Q^{-1}$. Without letting go of the ends of the rope, heave ho! Pull it tight. If the entire rope comes loose, then $P \cdot Q \cdot P^{-1} \cdot Q^{-1} \sim 1$. When you try it, you will see that this doesn't happen. Therefore, $P \cdot Q \cdot P^{-1} \cdot Q^{-1} \not\sim 1$. This means that $P \cdot Q \not\sim Q \cdot P$, which shows that $\pi_1(\text{Plane} - \{\text{two points}\}; O)$ is not commutative. Actually, the fundamental group is known to be the free group with two generators. (See Figure 6.3.)

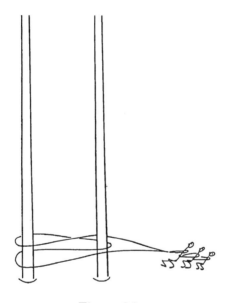

Figure 6.3.

Example n. Similarly, $\pi_1(\text{Plane} - \{n \text{ points}\}; O) \simeq$ the free group with n generators. If $n > 1$, this group is not abelian.

Example 1'. The fundamental group of an annulus. As in Example 1, the fundamental group is isomorphic to **Z**. Actually, this case can be reduced to Example 1. Let D' be the annulus with outer radius a and inner radius b. When we use polar coordinates (r, θ) to represent points in D', the correspondence φ given by

$$D' \ni (r, \theta) \overset{\varphi}{\mapsto} \left(\frac{r - b}{a - r}, \theta \right) \in \text{ the plane} - \{\text{the origin}\}$$

is a one-to-one and onto map from D' to (Plane $-$ {origin}) $= D$. Since both φ and φ^{-1} are continuous, D' and D are homeomorphic. (In general, two spaces are *homeomorphic* if there is a bijection ψ such that both ψ and ψ^{-1} are continuous.)

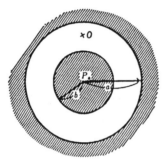

Figure 6.4.

We use $D' \approx D$ to denote this. In general, we have the following theorem.

Theorem 6.1.

If D and D' are homeomorphic, $\pi_1(D; O)$ and $\pi_1(D'; O')$ are isomorphic.

Here O and O' are points in D and D' respectively. Since the proof is easy, it is left as an exercise. (Just assign C to $\psi(C)$.)

Example 2'.   The region D of Figure 5.1 is homeomorphic to the plane minus two points. Therefore, this is reduced to Example 2.

Example n'. $D =$ an island with n lakes on it. Then D is homeomorphic to the plane minus n points.

The Seventh Week:
Examples of fundamental groups, continued

So far we have discussed fundamental groups of plane regions. Needless to say, we can define fundamental groups for surfaces in space.

Example 1. The surface in Figure 7.1 is called a torus. The surface of a doughnut and an inner tube (without considering the air inside) are tori.

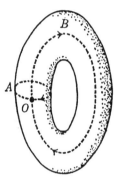

Figure 7.1.

Consider the fundamental group $\pi_1(T;O)$ of a torus T. Is it abelian?

Let us consider whether the equivalence classes $a = [A]$ and $b = [B]$ commute, where A and B are the curves in Figure 7.1. In order to see if $a \cdot b = b \cdot a$, we check whether $B^{-1}A^{-1}BA \sim 1$. If you tie a rope around a torus to represent $B^{-1}A^{-1}BA$, you can see that the rope will come loose without leaving the surface of the torus. This shows $B^{-1}A^{-1}BA \sim 1$. Hence $a \cdot b = b \cdot a$. (See Figure 7.2.)

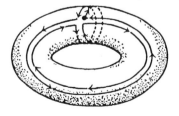

Figure 7.2.

It is known that every element of $\pi_1(T;O)$ can be written in the form $a^n \cdot b^m$ $(n, m = 0, \pm 1, \pm 2, \ldots)$; therefore, $\pi_1(T;O) \cong \mathbf{Z} \oplus \mathbf{Z}$. In particular, $\pi_1(T;O)$ is abelian.

Example 2.   The fundamental group of an inner tube for two, three, or four becomes more complicated. These groups are no longer abelian. (See Figure 7.3.)

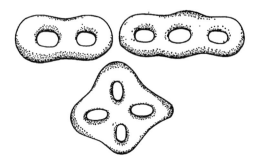

Figure 7.3.

We can define fundamental groups not only for two-dimensional surfaces, but also for three-dimensional regions in space. For example, a solid torus (that is, the essential part of a doughnut) or the air inside an inner tube. The fundamental group of a solid torus is isomorphic to **Z**.

Fundamental groups are defined for higher dimensional connected manifolds, as well as connected regions of higher dimensional Euclidean spaces.

In general, as long as you can define closed curves and continuous deformations of curves in a set, you can define a fundamental group.

Exercise.   Determine the fundamental group of the space obtained from $\mathbf{R}^3$ by deleting a circle S^1 and a line l through it, i.e., $D = \mathbf{R}^3 - (S^1 \cup l)$.

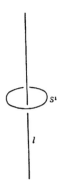

Figure 7.4.

Men Who Don't Realize
That Their Wives Have Been Interchanged

The Eighth Week:
Coverings

Let us start with an example. Take two distinct points O' and O in the plane. Let a and b be real numbers such that $0 < a < b$. Let D' be the set of points P whose distance from O' is between a and b. Similarly, let D be the region bounded by two circles of radii a and b and center O. See Figure 8.1.

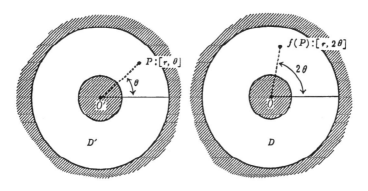

Figure 8.1.

We will use a polar cooordinate system having O' as the origin to express points in D', and another having O as the origin to express points in D.

We define a map f from D' to D by sending a point (r, θ) of D' to the point $(r, 2\theta)$ of D. Then f is a continuous map from D' to D. The point $f(P) = (r, 2\theta)$ moves in D as $P = (r, \theta)$ moves in D'. As P traces out the figure F in D', as in Figure 8.2, $f(P)$ traces out the figure in the right side of Figure 8.2. It is denoted by $f(F)$. As a point P moves in D' along a closed curve around O' once, $f(P)$ moves around O in D twice.

Let Q be a point in D, and consider $f^{-1}(Q)$. That is, find a point P in D' such that $f(P) = Q$. For example, if $Q = (c, 40°)$, then the point $P_1 = (c, 20°)$ satisfies $f(P_1) = Q$, but this is not the only solution of the equation $f(P) = Q$; $P_2 = (c, 200°)$ also satisfies $f(P) = Q$. Indeed, $f(c, 200°) = (c, 400° - 360°) = (c, 40°) = Q$. There are no other solutions to the equation $f(P) = (c, 40°)$ besides $P_1 = (c, 20°)$ and $P_2 = (c, 200°)$. That is, $f^{-1}(Q) = \{P_1, P_2\}$. In general, for every $Q = (r, \theta)$ of D,

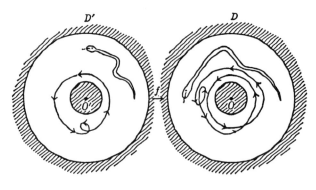

Figure 8.2.

$f^{-1}(Q)$ consists of two points $P_1 = (r, \frac{1}{2}\theta)$ and $P_2 = (r, \frac{1}{2}\theta + 180°)$. Thus f is a two-to-one map. See Figure 8.3.

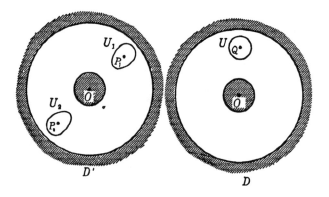

Figure 8.3.

Let U be a small neighborhood of $Q = (r, \theta)$ in D and consider its inverse image $f^{-1}(U)$. Since $f^{-1}(U)$ consists of points P in D' such that $f(P)$ is in U, it is a union of two small neighborhoods U_1 of $P = (r, \frac{1}{2}\theta)$ and U_2 around $P_2 = (r, \frac{1}{2}\theta + 180°)$. In other words, $f^{-1}(U) = U_1 \cup U_2$. (When U is a small disc around Q, U_1 and U_2 are egg-shaped sets around P_1 and P_2.) Of course, U_1 and U_2 have no intersection:

$$U_1 \cap U_2 = \emptyset$$

As P sweeps out all of U_1, $f(P)$ sweeps out all of U. If P_1 and P_2 are two points of U_1, $f(P_1)$ and $f(P_2)$ are two points of U: $P_1, P_2 \in U_1$, $P_1 \neq P_2 \Longrightarrow f(P_1) \neq f(P_2)$. That is, f is a two-to-one map of D' to D, but its restriction to U_1 is a one-to-one map of U_1 onto U. Moreover,

$$f|_{U_1} : U_1 \longrightarrow U \text{ is a homeomorphism onto } U$$
$$f|_{U_2} : U_2 \longrightarrow U \text{ is a homeomorphism onto } U.$$

Explanation: A continuous map of a domain or region U_1 onto U is a homeomorphism if it is one-to-one and its inverse map is continuous. Our f, as a map of D' to D, is globally two-to-one, so it is not a homeomorphism. However its restriction to U_1 is one-to-one, so we can define the inverse map $(f|_{U_1})^{-1} : U \longrightarrow U_1$, which is obviously continuous. Thus $f|_{U_1}$ is a homeomorphism.

Generalizing this example, we define the notion of a covering space.

Definition: Let D' and D be two connected regions in the plane. When a map $f : D' \to D$ satisfies the following conditions C1 and C2, we call f a *covering map* from D' to D. We also say that D' *covers* D by f, and that D' is a *covering space* (or surface) of D.

C1: f is a continuous surjection.

C2: If $Q \in D$, then $f^{-1}(Q)$ is a finite or countable set $\{P_1, P_2, \ldots, P_n, \ldots\}$. If U is a small neighborhood of Q in D, then

   (i) $f^{-1}(U)$ is the union of small neighborhoods U_1 of P_1, U_2 of P_2, ... That is, $f^{-1}(U) = U_1 \cup U_2 \cup \cdots \cup U_n \cup \cdots$.

   (ii) These small neighborhoods U_i do not intersect each other: $U_i \cap U_j = \emptyset$ if $i \neq j$.

   (iii) $f|_{U_i}$ is a homeomorphism onto U.

Let me paraphrase the latter half of C2: by taking a small enough neighborhood U of Q, we can always write $f^{-1}(U)$ as the union of $U_1, U_2, \ldots$ that satisfy the conditions above.

A continuous surjective map f is called a *covering map* if we can find such a U for each point Q of D.

Example 1. Let D be the surface of a right circular cylinder with radius 1 cm and height 10 cm. Let D' be an infinite horizontal strip of width 10 cm in the xy-plane: $D' = \{(x, y) \mid 0 < y < 10\}$.

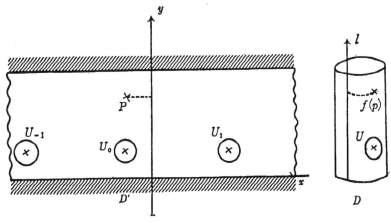

Figure 8.4.

Choose a vertical line on the cylinder. We define a map from D' to D as follows: Place D of top of D' so that the line l coincides with the y-axis. Roll (*goro-goro-go**) the cylinder over the strip D'. Roll it in both the positive and negative directions. If you mark a point P on D' in ink, then this will produce a mark Q on D. We let f be the map assigning to P the ink blot Q. Then f is a covering map from D' to D.

Example 2.   Take a region D in the plane, as in Figure 8.5.

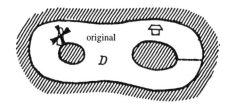

Figure 8.5.

Include in the region a dotted line connecting one lake to the sea. Take two elastic transparent sheets and follow these instructions:

1. Place one sheet over D and trace D on the sheet.

2. Cut it out.

1′. Put the other sheet over D and trace.

2′. Cut it out.

Call these two rubber shapes copy 1 of D and copy 2 of D. Also trace the dotted line on copy 1 and copy 2.

 3. Cut copy 1 and copy 2 along the dotted lines.

Name the segments at the edges of the cut as shown in Figure 8.6.

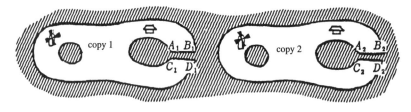

Figure 8.6.

 4. Turn copy 2 around. Open the cuts wider as in Figure 8.7.

---

* The sound of a large rolling object, such as a steam roller.

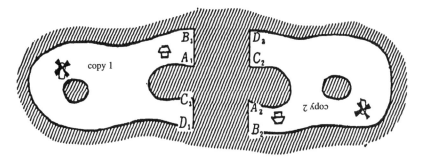

Figure 8.7.

5. Join copy 1 to copy 2 by grafting $\overline{A_1B_1}$ to $\overline{C_2D_2}$ and $\overline{A_2B_2}$ to $\overline{C_1D_1}$. You obtain a connected rubber shape D'. See Figures 8.8 and 8.9.

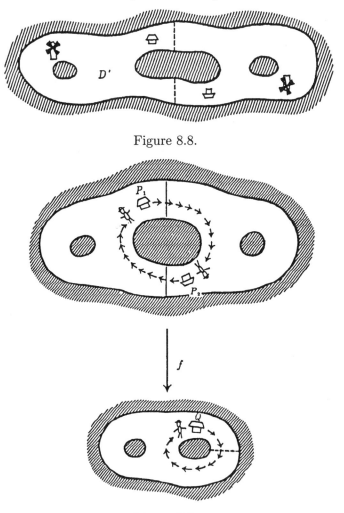

Figure 8.8.

Figure 8.9.

A map f from D' to the original D can be defined as follows: for each point P in D', take the point Q of D which was underneath P in Step 1 or 1'. If we assign Q to the point P, then f is a covering map from D' to D.

Let us trace everything in D onto the copies in Step 1 and 1'. So if there is a house at point Q in D, then there are houses at points P_1 and P_2 of D'. If there is a windmill in D, then there are two windmills in D' turning in the wind. If Mr. G lives in the house Q, then Mr. G_1 and Mr. G_2 live in the houses P_1 and P_2, respectively. Let's call G_1 and G_2 shadows of G. When Mr. G strolls around his house Q, Mr. G_1 must stroll around his house P_1 and Mr. G_2 must stroll around his house P_2. Suppose that Mr. G goes out for an excursion around the lake, as in Figure 8.9. Then Mr. G_1 must leave his house P_1 and walk along the lake, but the house he reaches is P_2. At the same time, the shadow Mr. G_2 reaches the house P_1 from P_2, walking along the lakeshore. However, the wives awaiting them at P_1 and P_2 have the same features, so Mr. G_1 and Mr. G_2 will never realize that they have come home to the wrong houses. The wives will never realize that the wrong husbands came home. Therefore, everything is consistent in this shadow world. The only possible crisis in this world is if Mr. G_1 meets Mr. G_2. Is there any such moment? In fact, it will be shown that Mr. G_1 can never meet Mr. G_2.

Does this fable lead you to deep but useless considerations on the identity of things? Or on the courage of mathematicians who identify two things which are only equivalent? But I don't like philosophy. Let's go back to mathematics.

The Ninth Week:
Covering surfaces and fundamental groups

Let $f : D' \longrightarrow D$ be a covering. Since f is a continuous map, when a point P in D' moves continuously over a figure F of D', $f(P) = Q$ moves continuously in D. Let $f(F)$ be the figure in D which Q traces out. In particular, if P traces a curve C in D', then the trace $f(C)$ of the point $f(P) = Q$ is again a curve in D. If C is a closed curve, then $f(C)$ is also a closed curve.

Of course,

Formula 9.1. $f(C_1 \cdot C_2) = f(C_1)f(C_2)$

holds.

If two curves A and B are homotopic in D', then $f(A)$ and $f(B)$ are homotopic in D. (See Figure 9.1.)

Formula 9.2. $A \sim B \Longrightarrow f(A) \sim f(B)$

Fix a point O' in D' and let $O = f(O')$. Consider $\pi_1(D'; O')$ and $\pi_1(D; O)$. For every element a of $\pi_1(D'; O')$, take a closed curve A such that $a = [A]$. Then $f(A)$ is a closed curve from O to O. Thus it defines a homotopy class $\alpha \in \pi_1(D; O)$ which contains $f(A)$.

Since α depends only on the homotopy class $a \in \pi_1(D'; O')$ and does not depend on the representative A (by Formula 9,2), let us denote α by $f_*(a)$. We denote by f_* the map assigning $f_*(a)$ in $\pi_1(D; O)$ to the element a of $\pi_1(D'; O')$. Formula 9.1 shows that f_* is a homomorphism from $\pi_1(D'; O')$ into $\pi_1(D; O)$. Thus the image $f_*(\pi_1(D'; O'))$ of f_* is a subgroup of $\pi_1(D; O)$.

Theorem 9.1. The covering $f : D' \longrightarrow D$ defines a subgroup $f_*(\pi_1(D'; O'))$ of $\pi_1(D; O)$. Conversely, for every subgroup γ of $\pi_1(D; O)$, there exists a covering $f : D' \longrightarrow D$ such that $\gamma = f_*(\pi_1(D'; O'))$.

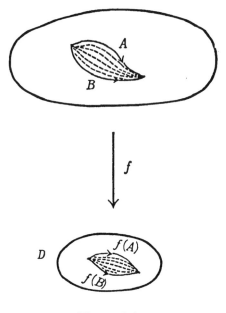

Figure 9.1.

The proof of this theorem will be postponed until next week. Then we will also see that f_* is injective. Therefore, the subgroup $f_*(\pi_1(D'; O'))$ is isomorphic to $\pi_1(D'; O')$.

The Tenth Week:
Covering surfaces and fundamental groups, continued

Let's recall what a covering map is:

C1: A continuous surjective map f from D' to D.

C2: For every point Q of D, $f^{-1}(Q)$ consists of finitely or countably many points $\{P_1, P_2, P_3, \ldots\}$ of D'. Also, there exists a small neighborhood U around each point Q such that there is a small neighborhood V_i around each P_i satisfying

(i) $f^{-1}(U) = V_1 \cup V_2 \cup V_3 \cup \cdots$, $P_i \in V_i$.

(ii) $V_i \cap V_j = \emptyset$ for $i \neq j$.

(iii) The restriction of f to V_i is a homeomorphism from V_i to U. That is, $f|_{V_i}$ is injective and surjective and $(f|_{V_i})^{-1}$ is continuous. See Figure 10.1.

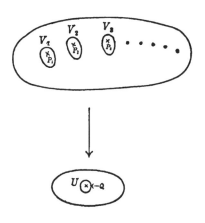

Figure 10.1.

Such a neighborhood U of D is called a *copiable neighborhood*. Each of the V_1, $V_2, \ldots$ is called a *copy* of U around P_1, $P_2, \ldots$.

Let $f : D' \longrightarrow D$ be a covering. For a curve C' in D', $f(C')$ is again a curve in D. $C = f(C')$ is called the *projection* of C'. We also say that C' covers C. See Figure 10.2.

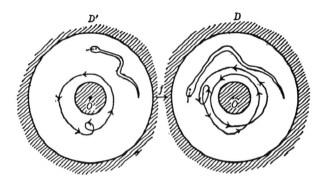

Figure 10.2.

Conversely, we can construct a curve C' in D' from a curve C in D by the following procedure. For the initial point Q of C, choose a point from $f^{-1}(Q) = \{P_1, P_2, \ldots\}$, say P_3.

(i)   Take a copiable closed neighborhood U around Q and consider a copy V_3 around P_3. Since the restriction of f to V_3, $f|_{V_3} : V_3 \longrightarrow U$, is a homeomorphism, its inverse map $(f|_{V_3})^{-1}$ maps U onto V_3 continuously. Therefore the segment $C \cap U = C_0$ of the curve C in U goes to a curve $C'_{3,0}$ in V_3 via $(f|_{V_3})^{-1}$. The initial point of $C'_{3,0}$ is, of course, P_3. Let $P_{3,1}$ be the endpoint of $C'_{3,0}$, and let $f(P_{3,1}) = Q_1$. Then Q_1 is the endpoint of $C \cap U = C_0$. See Figure 10.3-1. Of course, C_0 is the projection of $C'_{3,0}$.

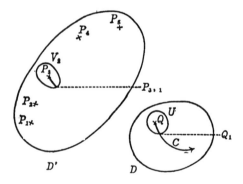

Figure 10.3-1.

(ii) Choose a copiable closed neighborhood U_1 around Q_1 and take a copy $V_{3,1}$ of U_1 around $P_{3,1}$. Map the segment $C \cap U_1 = C_1$ of the curve C contained in U_1 into $V_{3,1}$ by $(f|_{V_{3,1}})^{-1}$. It intersects $C'_{3,0}$ and extends it a little. See Figure 10.3-2.

We call the extension $C'_{3,1}$. Let $P_{3,2}$ be the endpoint of $C'_{3,1}$ and let $f(P_{3,2}) = Q_2$.

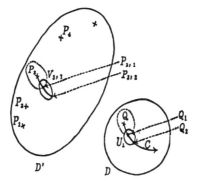

Figure 10.3-2.

Then $C_0 \cup C_1$ is nothing but the projection $f(C'_{3,0} \cup C'_{3,1})$ of $C'_{3,0} \cup C'_{3,1}$.

(iii) Choose a copiable closed neighborhood U_2 around Q_2 and take a copy $V_{3,2}$ of U_2 around $P_{3,2}$. Map the segment $C \cap U_2 = C_2$ of the curve C contained in U_2 into $V_{3,2}$ by $(f|_{V_{3,2}})^{-1}$. It intersects $C'_{3,1}$ and extends it a little. See Figure 10.3-3.

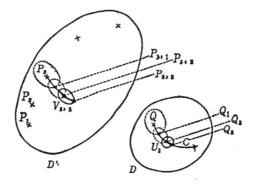

Figure 10.3-3.

We call the extension $C'_{3,2}$. Let $P_{3,3}$ be the endpoint of $C'_{3,2}$ and let $f(P_{3,3}) = Q_3$. Then $C_0 \cup C_1 \cup C_2$ is the projection $f(C'_{3,0} \cup C'_{3,1} \cup C'_{3,2})$ of $C'_{3,0} \cup C'_{3,1} \cup C'_{3,2}$.

(iv) Choose a copiable closed neighborhood U_3 around Q_3 and take a copy $V_{3,3}$ of U_3 around $P_{3,3}$. Map the segment $C \cap U_3 = C_3$ of the curve C contained in U_3 into $V_{3,3}$ by $(f|_{V_{3,3}})^{-1}$. It intersects $C'_{3,2}$ and extends it a little. We call the extension $C'_{3,3}$. Let $P_{3,4}$ be the endpoint of $C'_{3,3}$ and let $f(P_{3,4}) = Q_4$. It is the endpoint of the curve C_3. Needless to say,

$$f(C'_{3,0} \cup C'_{3,1} \cup C'_{3,2} \cup C'_{3,3})$$

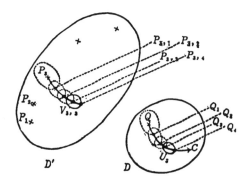

Figure 10.3-4.

fills up the curve C between Q and Q_4. See Figure 10.3-4.

Did you really bother to read all of this? Did you follow this explanation to the end?

By a successive application of these procedures, the arc $C'_{3,0}$ starting at P_3 in the covering space D' will be extended bit by bit (such as $C'_{3,0} \cup C'_{3,1}$; $C'_{3,0} \cup C'_{3,1} \cup C'_{3,2}$; $C'_{3,0} \cup C'_{3,1} \cup C'_{3,2} \cup C'_{3,3}$; $C'_{3,0} \cup C'_{3,1} \cup C'_{3,2} \cup C'_{3,3} \cup C'_{3,4}$; ...) and eventually

$$\bigcup_{j=0}^{n} C'_{3,j}$$

will cover the whole curve C. That is, $f(\bigcup_{j=0}^{n} C'_{3,j}) = C$. (The reason for this is that, since C contains both endpoints, it is compact. See any topology book.)

So if we let $C'_3 = \bigcup_{j=0}^{n} C'_{3,j}$, then C'_3 is a curve in D' starting at P_3 whose projection is C. Since the curve C'_3 is determined by the curve C and the initial point P_3 in D', we call it the *lift* of C with initial point P_3. Because f is a homeomorphism on each copy of a copiable neighborhood, the lift is unique on each small neighborhood. Since the lift is unique at each step, and the small segments overlap, the lift of the entire curve is unique. In other words, the lift of a curve C depends only on C and the initial point chosen for the lift.

In the procedure, we chose P_3 in $f^{-1}(Q) = \{P_1, P_2, P_3, \ldots\}$ in order to construct the curve C'_3, but of course we can do the same procedure for any other points P_1 or P_2 or P_3 .... Let C'_1 be the lift of C starting at P_1, C'_2 be the lift of C starting at P_2.... The number of lifts of the curve C is equal to the cardinality of $f^{-1}(Q)$, where Q is the initial point of C. Let M_1 be the endpoint of C'_1, M_2 be the endpoint of C'_2, ... It is obvious that the points $M_1, M_2, \ldots$ cover the endpoint N of C. That is, $f^{-1}(N) = \{M_1, M_2, \ldots\}$.

Take a point R on C. Every point $S_1, S_2, \ldots$ of the inverse image $f^{-1}(R) = \{S_1, S_2, \ldots\}$ is on a curve $C'_1, C'_2, \ldots$. See Figure 10.4.
We can renumber the indices i so that S_i is on C'_i. As R moves along the curve C, each of the points $S_1, S_2, \ldots$ moves along $C'_1, C'_2, \ldots$. We call R the original and S_i

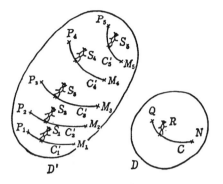

Figure 10.4.

the shadow of R. As we noted, when the original R moves along C, its shadows $S_1, S_2, \ldots$ move along the curves $C'_1, C'_2, \ldots$ in D'. But can they crash into each other in D'? It doesn't happen. See Figure 10.5.

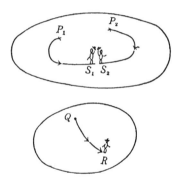

Figure 10.5.

Proof: Suppose S_1 crashes into S_2 when R comes to a point R_0 in C. Let S_0 be the site of the accident. That is, $S_1 = S_2 = S_0$ when $R = R_0$. Let U be a copiable neighborhood of R_0 and let V_0 be a copy of U around S_0. Then it is obvious from Figure 10.6 that $f|_{V_0}$ is no longer one-to-one. This contradicts the definition of a covering. Therefore, S_i will never crash into S_j. Q.E.D.

Problem: For a covering $f : D' \longrightarrow D$, let n ($1 \leq n \leq \infty$) be the cardinality of the inverse image $f^{-1}(Q) = \{P_1, P_2, P_3, \ldots\}$ of a point Q in D. Show that n doesn't depend on the choice of Q. This number is called the *degree* of the covering f and is denoted by $\deg(f)$ or $n = [D' : D]$. When $n \neq \infty$, we say that D' is a finite covering of D. As we announced last week, now we can prove that f_* is injective by using lifting.

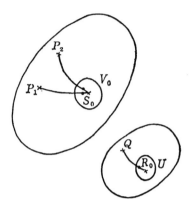

Figure 10.6.

Preparation Theorem. Let C_0 and C_1 be two curves in D which are homotopic. So, in particular,

$$\text{initial point of } C_0 = \text{ initial point of } C_1.$$

We call this point O. Let $f : D' \longrightarrow D$ be a covering and $O' \in D'$ be a point such that $f(O') = O$. Let C_0' and C_1' be the lifts of C_0 and C_1 starting at O'. Of course we have the equality

$$O' = \text{ initial point of } C_0' = \text{ initial point of } C_1'.$$

Then we have
$$\text{terminal point of } C_0' = \text{ terminal point of } C_1'$$

and
$$C_0' \sim C_1'.$$

Proof: Since $C_0 \sim C_1$, we can deform C_0 continuously into C_1. We assume that the deformation starts at time $t = 0$ and ends at time $t = 1$. Let $C_{\frac{1}{n}}, C_{\frac{2}{n}}, C_{\frac{3}{n}}, \ldots,$ $C_{\frac{n-1}{n}}$ be the curve at the intermediate stages of deformation at times $\frac{1}{n}, \frac{2}{n}, \ldots, \frac{n-1}{n}$. See Figure 10.7.

Next we divide each of these curves into m parts and make a net by connecting the division points as in Figure 10.8.

If we make n and m large enough so that the mesh is small, then we can assume that each square of the net is contained in a copiable neighborhood which is homeomorphic to a disc. Now we make a change in each step of the deformation of $C_{\frac{k}{n}}$

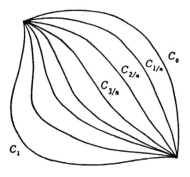

Figure 10.7.

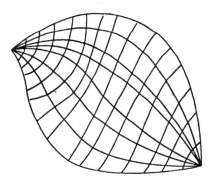

Figure 10.8.

to $C_{\frac{k+1}{n}}$ in the following way. In Figure 10.9, we follow the curve $C_{\frac{k+1}{n}}$ until the j-th horizontal stitch of the net, then follow the vertical stitch in the net to the j-th horizontal stitch of $C_{\frac{k}{n}}$, then follow $C_{\frac{k}{n}}$ to the end.

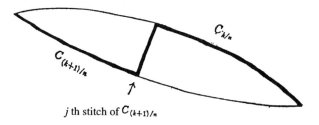

j th stitch of $C_{(k+1)/n}$

Figure 10.9.

Call this path $E_{k,j}$. As we can see in Figure 10.10, $E_{k,j}$ differs from $E_{k,j+1}$ on a very small piece (within a small square of the net), and this small square is contained in a copiable disc U. Therefore, $E_{k,j}$ can be continuously deformed to $E_{k,j+1}$ within U as shown in Figure 10.10.

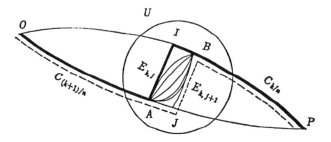

Figure 10.10.

Let us denote this deformation by $E_{k,j} \longrightarrow E_{k,j+1}$. Thus we obtain a sequence of continuous deformations

$$C_{\frac{k}{n}} = E_{k,0} \longrightarrow E_{k,1} \longrightarrow E_{k,2} \longrightarrow \ldots \longrightarrow E_{k,m} = C_{\frac{k+1}{n}}$$

We can apply this sequence for each $k = 0, 1, 2, \ldots, n-1$ to obtain a continuous deformation of C_0 to C_1. Thus the continuous deformation $C_0 \longrightarrow C_1$ is decomposed into nm small deformations $E_{k,j} \longrightarrow E_{k,j+1}$. In order to establish the theorem, we must show the following:

$$E'_{k,j} \sim E'_{k,j+1} \text{ and } P'_{k,j} = P'_{k,j+1}$$

for each k and j, where $E'_{k,j}$ is the lift of $E_{k,j}$ starting at O' and $P'_{k,j}$ is the endpoint of $E_{k,j}$. However, $E'_{k,j} \sim E'_{k,j+1}$ and $P'_{k,j} = P'_{k,j+1}$ are almost trivial, because each small square of the net is contained in a copiable neighborhood U. It is easily understood if you remember the construction of lifting. Even so, let me give you the details.

Let A be the point where $C_{\frac{k+1}{n}}$ and $E_{k,j}$ diverge, in Figure 10.10, and let B be the point where $E_{k,j+1}$ and $C_{\frac{k}{n}}$ converge. Since $E_{k,j}$ coincides with $E_{k,j+1}$ between O and A, their lifts coincide between O' and A' because of the uniqueness of lifting. Here A' is a point of D' which projects to A. In order to extend the lift of $E_{k,j}$ beyond the point A', we take a copiable neighborhood U of A and its copy U' around A' (as we did at the beginning of today's lecture) and map the arc $\overline{AB}$ into U' via the homeomorphism $(f|_{U'})^{-1} : U \to U'$. Compare the lifts $E'_{j,k}$ and $E'_{j,k+1}$ of the pieces of $E_{j,k}$ and $E_{j,k+1}$ which start at A and end at B. We have already seen that both these lifts start at A', and since $(f|_{U'}) : U' \to U$ is one-to-one, the endpoints are also the same. Call the common endpoint B'. The lift $E'_{k,j}$ of $E_{k,j}$ and the lift $E'_{k,j+1}$ of $E_{k,j+1}$ meet at the point B'. Because of the uniqueness of the lift of the arc $\overline{BP}$ we can show that $E'_{k,j}$ coincides with $E'_{k,j+1}$ after the point B', and, in particular, the endpoint $P'_{k,j}$ coincides with $P'_{k,j+1}$. Since the difference between $E'_{k,j}$ and $E'_{k,j+1}$ is contained in the small disc U', $E'_{k,j} \sim E'_{k,j+1}$. This completes the proof of the theorem.

Now we have

Theorem 10.1. Let $f : D' \longrightarrow D$ be a covering of D, O' and O points of D' and D, respectively, such that $f(O') = O$. Then the homeomorphism $f_* : \pi_1(D'; O') \longrightarrow \pi_1(D; O)$ induced by f is an injection.

Proof: Let γ' be an element of $\pi_1(D'; O')$ such that $f_*(\gamma') = 1$. That is, for a closed curve C' belonging to γ', the curve $C = f(C')$ is null-homotopic in D. Therefore, the lift C' of C is also null-homotopic by the Preparation Theorem. Thus $\gamma' = 1$. Therefore the kernel of f_* is $\{1\}$, hence f_* is an injection.

The Eleventh Week:
The Group of Covering Transformations

Let $f : D' \longrightarrow D$ be a covering. We say that points P_1 and P_2 of D' are *conjugate* if $f(P_1) = f(P_2)$. The number of points which are conjugate to P_1 (including P_1 itself) is equal to $n = \deg(f)$. Similarly, we say that curves C'_1 and C'_2 are *conjugate* if they are lifts of the same curve. In order to construct a curve starting at P_2 which is conjugate to a curve C'_1 starting at P_1, first we project C'_1 to make a curve C. Then take the lift of C which starts at P_2. The number of curves conjugate to C'_1 (including C'_1 itself) in D' is also equal to $n = \deg(f)$.

The covering $f : D' \longrightarrow D$ is said to be a *Galois* or *normal* covering of D if every conjugate of a closed curve C' in D' is again a closed curve. Figure 11.1 shows a covering which is not Galois. Among the lifts of C, C'_1 and C'_2 are not closed curves, but C'_3 is.

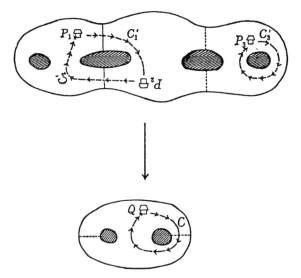

Figure 11.1. Covering which is not Galois.

Definition: Let $f : D' \longrightarrow D$ be a covering. A homeomorphism σ of D' onto itself is said to be a *covering transformation* of $f : D' \longrightarrow D$ if

$$f(\sigma(P)) = f(P)$$

for every $P \in D'$. The set of all covering transformations forms a group which is denoted by $\Gamma(D' \xrightarrow{f} D)$. We call the group the *covering transformation group* of $f : D' \longrightarrow D$.

It is obvious that $\Gamma(D' \xrightarrow{f} D)$ is a group. Let $\sigma \in \Gamma(D' \xrightarrow{f} D)$. If $\sigma(P_1) = P_2$, then $f(P_2) = f(\sigma(P_1)) = f(P_1)$. Hence, P_1 and P_2 are conjugate. Similarly, let B_1' be an arbitrary curve in D', and set $\sigma(B_1') = B_2'$. Then B_1' and B_2' are conjugate. Let S_1 be an arbitrary point of D' and C_1' a curve in D' which starts at P_1 and ends at S_1. Let C_2' be a conjugate of C_1' starting at P_2 and let S_2 be its endpoint. Since $\sigma(P_1) = P_2$, $\sigma(C_1')$ is a conjugate curve of C_1' starting at P_2. Therefore it coincides with $C_2' : \sigma(C_1') = C_2'$. Hence $\sigma(S_1) = S_2$.

This argument tells us that the covering transformation σ is completely determined by the image $\sigma(P_1) = P_2$ of a single point P_1. See Figure 11.2.

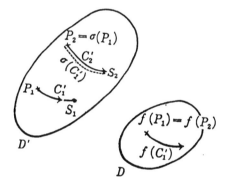

Figure 11.2.

To summarize,

Proposition 11.1. Let σ be a covering transformation. If $\sigma(P_1) = P_2$, then P_1 and P_2 are conjugate. In addition, for every pair (P_1, P_2) of points in D' which are conjugate, there is at most one covering transformation σ such that $\sigma(P_1) = P_2$.

Theorem 11.1. Let $f : D' \longrightarrow D$ be a Galois covering. Let P_1 and P_2 be an arbitrary pair of conjugate points in D'. Then there exists a unique covering transformation σ such that $\sigma(P_1) = P_2$. We denote this by $\sigma(P_1; P_2)$. Let $\{P_1, P_2, P_3, \ldots\}$ be the set of all points which are conjugate to P_1. Then the set of all covering transformations of $f : D' \longrightarrow D$ is

$$\{\sigma(P_1; P_1) = \text{ identity}, \ \sigma(P_1; P_2), \ \sigma(P_1; P_3), \ldots\}$$

In particular, the number of covering transformations of $f : D' \longrightarrow D$ is exactly $n = \deg(f)$.

Proof: Let P_1 and P_2 be a pair of conjugate points in D'. For an arbitrary point Q_1 in D', we draw a curve C_1' in D' which connects P_1 and Q_1, and let C_2' be a conjugate of C_1' starting at P_2. Let Q_2 be the endpoint of C_2'. Let us prove that this point Q_2 is completely determined by Q_1 and does not depend on the choice of C_1'. Let C_1'' be another curve connecting P_1 and Q_1. Let C_2'' be the conjugate curve of C_1'' starting at P_2, and let Q_2'' be its endpoint. Since $f : D' \longrightarrow D$ is Galois and $C_1'^{-1} \cdot C_1''$ is a closed curve, its conjugate $C_2'^{-1} \cdot C_2''$ is also a closed curve. Therefore $Q_2'' = Q_2$. Thus Q_2 is determined by Q_1. See Figure 11.3.

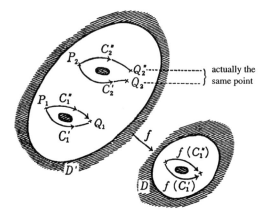

Figure 11.3.

Now let σ be a map assigning to every Q_1 the point Q_2 thus obtained. It is easy to see that this σ is a covering transformation satisfying $\sigma(P_1) = P_2$.   Q.E.D.

Theorem 11.2. If $f : D' \longrightarrow D$ is a Galois covering, then

(1) $f_*(\pi_1(D'; O'))$ is a normal subgroup of $\pi_1(D; O)$.

(2) $\pi_1(D; O)/f_*(\pi_1(D'; O')) \cong \Gamma(D' \xrightarrow{f} D)$, where O' is a point in D' such that $O = f(O')$.

Sketch of proof: Let x be an arbitrary element of $\pi_1(D; O)$. Assume $x = [A]$, where A is a closed curve in D from O to O. Let A' be the lift of A starting at O'. A' may not be a closed curve. We denote by $P(x)$ its endpoint. It can be shown that $P(x)$ depends only on x and does not depend on the representative A of the class x. It is clear that $P(x)$ is conjugate to O'. However, it may not be equal to O'. Let us consider $\sigma(O'; P(x)) \in \Gamma(D' \xrightarrow{f} D)$. For simplicity, we write $\psi(x)$ for $\sigma(O'; P(x))$. If we define a map ψ which assigns to each x in $\pi_1(D; O)$ the group element $\psi(x) \in \Gamma(D' \xrightarrow{f} D)$ thus obtained, then we have the following:

(i) ψ is a homomorphism. Namely, $\psi(x \cdot y) = \psi(x) \cdot \psi(y)$.

(ii) ψ is surjective. That is, $\psi(\pi_1(D; O)) = \Gamma(D' \xrightarrow{f} D)$.

(iii) The kernel of ψ is $f_*(\pi_1(D'; O'))$.

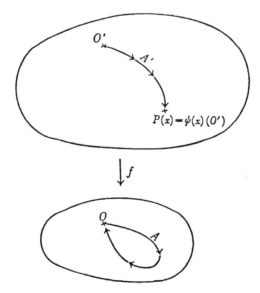

Figure 11.4.

The proof of (i) is sketched in Figure 11.5.

Proof that $\psi(x \cdot y) = \psi(x) \cdot \psi(y)$: We apply the method of constructing σ used in

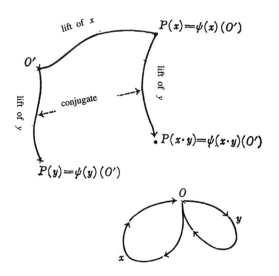

Figure 11.5.

the previous theorem to $\sigma = \psi(x)$ and $Q = P(y)$. We obtain $P(x \cdot y) = \psi(x) \cdot P(y)$. Thus, $\psi(x \cdot y)(O') = \psi(x) \cdot \psi(y)(O')$. Hence, $\psi(x \cdot y) = \psi(x) \cdot \psi(y)$.

The proof of (ii) is left to the reader.

Proof of (iii): Since $\psi(x) = 1 \iff P(x) = O' \iff A'$ determines an element x' of $\pi_1(D'; O')$, it is obvious that $f(x') = x$. Therefore $x \in f_*(\pi_1(D'; O'))$. Conversely, if $x \in f_*(\pi_1(D'; O'))$ with $x = A$, then the lift A' is a closed curve and we have $P(x) = O'$.

Returning to the proof of the theorem, since $f_*(\pi_1(D'; O'))$ is the kernel of a homomorphism, it is a normal subgroup of $\pi_1(D; O)$. By the main homomorphism theorem of group theory,

$$\pi_1(D; O)/f_*(\pi_1(D'; O')) \cong \Gamma(D' \xrightarrow{f} D)$$

Q.E.D.

Everyone has a tail

The Twelfth Week:
The Universal Covering Space

When I was a schoolboy, I was seized by the idea that everyone has an invisible tail at his back. This tail is very light, a million miles long, very thin, and trails behind as we walk, from birth to death. The tail extends far away, beyond our world, and originates in the Land of the Dead. When we walk, the tail is reeled out from its source. When we have pulled out the entire tail, we die.

The strategy for a long life, therefore, is to try to avoid pulling out your tail. In order to do this, walk null-homotopic paths as much as possible. When you return, follow the same route. Never go through a tunnel. (My ideal world was a universal covering space of the real world (if there is such a thing).) My idea completely obsessed me; I was trapped by my own fantasy. I even came to loathe people who crossed my path behind me.

Figure 12.1.

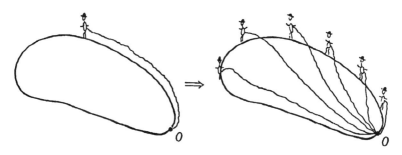

Figure 12.2.

As we noted before, the notions of fundamental group and covering $f : D' \longrightarrow D$ can be defined not only for plane regions, but also for surfaces in space, domains in space, higher dimensional manifolds, and even in abstract spaces in which the notions of neighborhoods, closed curves, and their continuous deformations are defined. What we are going to deal with here are two-dimensional manifolds.

What is a manifold? In order to define a manifold, we need some ideas from topology. In the beginning of this lecture series, I required that you study a few pages of a topology textbook. Have you done this? In what follows, we assume that you know the language of topology.

A two-dimensional manifold is a topological space such that each point has a neighborhood which is homeomorphic to a plane region. In particular, a plane region is itself a two-dimensional manifold. Let $\tilde{D}$ and D be two-dimensional manifolds and $f : \tilde{D} \longrightarrow D$ be a covering. We assume that $\pi_1(\tilde{D}) = \{1\}$; namely, $\tilde{D}$ is simply connected. When this happens, we say that $\tilde{D}$ is a universal covering surface of D. In summary,

Definition: A universal covering surface is a simply connected covering surface. (For example, Figure 8.4 in the 8th week shows a universal covering surface.)

Now the question is, is there a universal covering surface for a given region D? Answer: certainly, there is. Our purpose this week is to prove this. We prove it by constructing $\tilde{D}$ using the following iterated construction. However, before going into detail, let us describe the construction intuitively. Let D be a plane region, such as the annulus in Figure 12.3. Choose a point O in D. We consider "points with tail" $\tilde{P}$ instead of the usual points in D. We assume that the end of the tail is fixed at the base point O.

We also imagine that the tail is made of rubber string, and we identify two points with tail if they have homotopic tails, as in Figure 12.3. Of course, the two points with tail in Figure 12.4 are distinct, because the tail C_1 is not homotopic to the tail C_3. We denote by $\tilde{D}$ the set of all such points with tail $\tilde{p}$. Then $\tilde{D}$ is the universal covering surface of D.

In the example in Figure 12.4, there are infinitely many points with tail which are the same point if we remove the tail. They are classified by the number of times

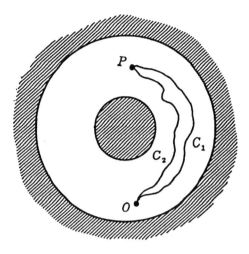

Figure 12.3.

that the tail winds around the pond. Therefore the universal covering surface of the annulus of Figure 12.4 is an infinitely long spiral staircase.

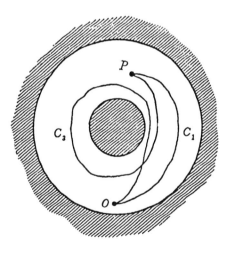

Figure 12.4.

Let us give a rigorous construction of the universal covering surface of D described intuitively above. We fix a point O in D and call it the origin. We denote by $V(D; O)$ the set of all curves in D starting at O (not neccessarily closed curves). As we said in the 4th week, a curve has an orientation, end points, and finite length. We continue to use this convention. Therefore, every element C of $V(D; O)$ has an initial point, which is, by definition, O, and an endpoint. Let us denote the

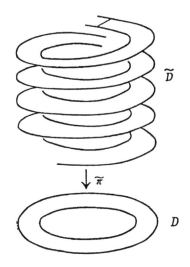

Figure 12.5.

endpoint of C by $e(C)$. In the 4th week, we defined $W(D)$, the set of all curves in D. $V(D;O)$ is the subset of $W(D)$ consisting of curves starting at O. In the 5th week, we considered the set $W(D;O)$ of closed curves starting and ending at O. This is the subset of $V(D;O)$ such that $e(C) = O$. Namely,

$$W(D) \supset V(D;O) \supset W(D;O).$$

We said that two curves C_1 and C_2 are homotopic ($C_1 \sim C_2$) if the initial point of C_1 is the initial point of C_2, the endpoint of C_1 is the endpoint of C_2, and C_1 can be continuously deformed to C_2 without moving the endpoints. Therefore if $C_1 \sim C_2$ and $C_1 \in V(D;O)$, then $C_2 \in V(D;O)$ and $e(C_1) = e(C_2)$. Similarly, if $C_1 \sim C_2$ and $C_1 \in W(D;O)$, then $C_2 \in W(D;O)$. Let us denote by $\tilde{D}$ the quotient space of $V(D;O)$ with respect to the equivalence relation $\sim$: $\tilde{D} = V(D;O)/\sim$. Elements of $\tilde{D}$ are denoted by $\tilde{P}, \tilde{Q}, \ldots$, and we call them tailed points of D or "points" of $\tilde{D}$. Let ν be the natural projection of $V(D;O)$ onto $\tilde{D}$. (Recall the definition at the end of Week 2.) If a point $\tilde{P}$ of $\tilde{D}$ is an equivalence class containing a curve C, then $\tilde{P} = \nu(C)$. That is, $\tilde{P} \ni C \iff \tilde{P} = \nu(C)$. Of course, $\tilde{D}$ contains $\pi_1(D;O) = W(D;O)/\sim$:

$$\tilde{D} \supset \pi_1(D;O).$$

Although $\pi_1(D;O)$ is a group, $\tilde{D}$ is a set containing $\pi_1(D;O)$ but has no group structure. We denote by $\tilde{O}$ the unit element 1 of $\pi_1(D;O)$, considered as an element of $\tilde{D}$: $\tilde{O} = 1$. Why can't we use two different symbols for a single object if we want? $\tilde{O} = 1$ is the equivalence class of closed curves in $V(D;O)$ which are continuously contractible to the point O.

We can define a surjective map from $V(D;O)$ onto D by assigning the endpoint $e(C)$ to every curve C in $V(D;O)$. Since $e(C_1) = e(C_2)$ when $C_1 \sim C_2$, we can define a natural surjection π from $\tilde{D} = V(D;O)/\sim$ onto D. Namely, for every point $\tilde{P}$ of $\tilde{D}$, we define $\pi(\tilde{P})$ to be the endpoint $e(C)$ of a curve C belonging to the class $\tilde{P}$. If $\tilde{P} = \nu(C) = \nu(C')$, then $C \sim C'$ and hence $e(C) = e(C')$. Therefore $\pi(\tilde{P})$ depends only on $\tilde{P}$ and is independent of the choice of C in $\tilde{P}$. It is obvious through the construction that $\pi \circ \nu = e$. Clearly, $\pi(\tilde{O}) = O$.

$$V(D;O) \xrightarrow{\ \nu\ } \tilde{D} \ni \tilde{O}$$
$$e \searrow \qquad \downarrow \pi$$
$$D \ni O$$

Now our plan is to introduce a topology on $\tilde{D}$ which makes it a two-dimensional manifold. Let's define a neighborhood of a point $\tilde{P}$ of $\tilde{D}$ in the following way. Let $\pi(\tilde{P}) = P$. Then P is a point of D which is the endpoint of a curve C belonging to $\tilde{P}$. Since D is a plane region, we can consider a neighborhood $U(P)$ around the point P. For simplicity, we assume that $U(P)$ is a disc $U_\epsilon(P)$ of radius ϵ and center P. We denote by $\overrightarrow{PQ}$ the line segment in $U_\epsilon(P)$ beginning at P and ending at Q. Choose a curve C which belongs to the class $\tilde{P}$, and let $\tilde{Q}$ be the class $\nu(C \cdot \overrightarrow{PQ})$ of the product $C \cdot \overrightarrow{PQ}$: that is, $\tilde{Q}$ is the set of curves which are homotopic to the curve $C \cdot \overrightarrow{PQ}$. This curve $C \cdot \overrightarrow{PQ}$ starts at O, follows C to P, then follows the line segment to Q. It is obvious that $\tilde{Q}$ depends only on $\tilde{P}$ and the point Q of $U_\epsilon(P)$ and is independent of the choice of C. So we define the "ϵ-neighborhood" $\tilde{U}_\epsilon(\tilde{P})$ of the point $\tilde{P}$ by

$$\tilde{U}_\epsilon(\tilde{P}) = \{\tilde{Q} = \nu(C \cdot \overrightarrow{PQ}) \mid Q \in U_\epsilon(P)\}.$$

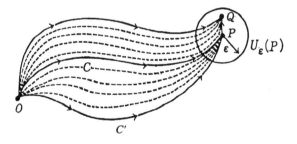

Figure 12.6.

That is, $\tilde{U}_\epsilon(\tilde{P})$ is the set of all $\tilde{Q} = \nu(C \cdot \overrightarrow{PQ})$ for all Q in the disc $U_\epsilon(P)$. It is a subset of $\tilde{D}$ which depends only on the class $\tilde{P}$ and the radius ϵ. Thus we have a system $\{\tilde{U}_\epsilon(\tilde{P})\}$ of ϵ-neighborhoods of various radii ϵ for each point $\tilde{P}$ of $\tilde{D}$. This system defines a topology on $\tilde{D}$. It is easy to see that $\tilde{D}$ is actually a two-dimensional manifold and that $\pi : \tilde{D} \longrightarrow D$ is a covering of D. Indeed, every point $\tilde{P}$ of $\tilde{D}$ has a neighborhood $\tilde{U}_\epsilon(\tilde{P})$ which is homeomorphic to $U_\epsilon(P)$, and π becomes a homeomorphism when restricted to $\tilde{U}_\epsilon(\tilde{P})$.

Finally, let's prove that $\tilde{D}$ is simply connected. Then we will know that $\tilde{D} \xrightarrow{\pi} D$ is a universal covering surface.

We choose an arbitrary closed curve $\tilde{C}$ in $\tilde{D}$ (at $\tilde{O}$), and show that it can be continuously deformed to $\tilde{O}$.

Suppose a moving point $\tilde{P}$ starts at $\tilde{O}$ when $t = 0$ and goes around $\tilde{C}$ once, and comes back to $\tilde{O}$ when $t = 1$. Let $\tilde{P}(t)$ be the position of $\tilde{P}$ at time t ($0 \le t \le 1$). Since $\tilde{C}$ is a closed curve, $\tilde{P}(1) = \tilde{O}$. Let $\pi(\tilde{P}) = P$, $\pi(\tilde{P}(t)) = P(t)$, and $\pi(\tilde{C}) = C$. Now C is a closed curve in D, starting and ending at O. Therefore, $P(1) = O$. The point $\tilde{P}(t)$, when considered as a "point with tail", is the point $P(t)$ with the part of C between O and $P(t)$ considered as its tail. Therefore, the image of $\tilde{P}(t)$ moving on $\tilde{C}$ can be considered as $P(t)$ moving on C, dragging its tail from O. In particular, $\tilde{P}(1)$, as a point with tail, is the point $P(1) = O$ with C as its tail. On the other hand, since $\tilde{O}$, as a point with tail, is the point O with a null-homotopic tail, $\tilde{P}(1) = \tilde{O}$ means that C is null-homotopic. By the Preparation Theorem in the 10th week, $\tilde{C}$ is also null-homotopic. Since an arbitrary closed curve $\tilde{C}$ in $\tilde{D}$ is null-homotopic, $\tilde{D}$ is simply connected.

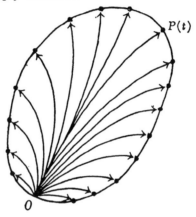

$P(t)$

0

Figure 12.7.

From the definition of the universal covering space, we have

Proposition 12.1. A universal covering $(\tilde{D}; \tilde{O}) \xrightarrow{\pi} (D; O)$ is a Galois covering.

Problem: Universality of *the* universal covering surface: Let $\tilde{D} \xrightarrow{\pi} D$ be the universal covering of D, $\tilde{O} \in \tilde{D}$, $O \in D$, and $\pi(\tilde{O}) = O$. Let $D' \xrightarrow{f} D$ be a covering, $O' \in D'$, and $f(O') = O$. Prove that there is a unique covering map $g : \tilde{D} \longrightarrow D'$ such that $g(\tilde{O}) = O'$ and $f \circ g = \pi$. (Hint): Let C be a curve in D, and $\tilde{P}$ the class C belongs to. Let C' be the lift of C in D' starting at O', and let P' be the endpoint of C'. Define g by assigning P' to $\tilde{P}$.

The Thirteenth Week:
The correspondence between coverings of $(D; O)$
and subgroups of $\pi_1(D; O)$

We first introduce a notation: $(D'; O') \xrightarrow{f} (D; O)$ means that $D' \xrightarrow{f} D$ is a covering and $f(O') = O$.

We saw in Weeks 10 and 11 that the homomorphism f_* induced by $(D'; O') \xrightarrow{f} (D; O)$ is injective and defines a subgroup $f_*(\pi_1(D'; O')) \subset \pi_1(D; O)$. This subgroup is isomorphic to $\pi_1(D'; O')$. Further, if $D' \xrightarrow{f} D$ is a Galois covering, then $f_*(\pi_1(D'; O'))$ is a normal subgroup, and the quotient group

$$\pi_1(D; O)/f_*(\pi_1(D'; O'))$$

is isomorphic to the group of covering transformations $\Gamma(D' \xrightarrow{f} D)$.

Let us consider the converse problem this week. That is, we will see that a subgroup Γ of $\pi_1(D; O)$ gives rise to a covering $(D'; O') \xrightarrow{f} (D; O)$ with the property $f_*(\pi_1(D'; O')) = \Gamma$.

First of all, if $\Gamma = \{1\}$, we can take $(\tilde{D}; \tilde{O})$ as $(D'; O')$. Then $f_*(\pi_1(\tilde{D}; \tilde{O})) = f_*(\{1\}) = \{1\}$. Therefore, by Theorem 11.2, the covering transformation group $\Gamma(\tilde{D} \xrightarrow{f} D)$ of $\tilde{D} \to D$ is isomorphic to $\pi_1(D; O) \cong \pi_1(D; O)/\{1\}$; in other words, $\Gamma(\tilde{D} \xrightarrow{f} D) \cong \pi_1(D; O)$.

Next, let Γ be a subgroup of $\pi_1(D; O)$. First construct the universal covering $(\tilde{D}; \tilde{O})$ of $(D; O)$. Since $\Gamma(\tilde{D} \xrightarrow{f} D) \cong \pi_1(D; O)$, we can consider Γ as a subgroup of $\Gamma(\tilde{D} \xrightarrow{f} D)$. We introduce an equivalence relation $\approx_\Gamma$ among points $\tilde{P}, \tilde{Q}, \ldots$ of $\tilde{D}$. We define $\tilde{P} \approx_\Gamma \tilde{Q}$ if there is a covering transformation γ such that $\gamma(\tilde{P}) = \tilde{Q}$. It is easy to see that $\approx_\Gamma$ is an equivalence relation. Then we can consider the quotient space $\tilde{D}/\approx_\Gamma$. According to our earlier notation, we would use $\tilde{D}/\Gamma$ to denote the quotient space, but the mathematical convention is to use $\Gamma \backslash \tilde{D}$ in this situation (a group acting on a space). It can be proved that $\Gamma \backslash \tilde{D}$ is a two-dimensional manifold and a covering space of D, and that the covering corresponds to the subgroup Γ of $\pi_1(D; O)$. The proof is left to the reader.

Now let us investigate when two coverings $(D';O') \xrightarrow{f'} (D;O)$ and $(D'';O'') \xrightarrow{f''}$ $(D;O)$ correspond to the same subgroup Γ of $\pi_1(D;O)$. For this purpose, we introduce the following equivalence relation among the coverings of $(D;O)$.

Definition: $(D';O') \xrightarrow{f'} (D;O)$ and $(D'';O'') \xrightarrow{f''} (D;O)$ are *equivalent coverings* of $(D;O)$ if there is a homeomorphism $g : D' \longrightarrow D''$ satisfying

(1) $g(O') = O''$

(2) $f'' \circ g = f'$

(Show that if such g exists, it is unique.)

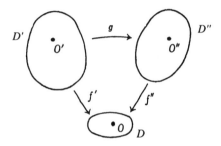

Figure 13.1.

Theorem 13.1. Let $(D';O') \xrightarrow{f'} (D;O)$ and $(D'';O'') \xrightarrow{f''} (D;O)$ be two coverings. These coverings are equivalent if and only if $f'_*(\pi_1(D';O')) = f''_*(\pi_1(D'';O''))$.

The proof is left to the reader. (In the original lecture series, the proofs of all the statements in this chapter were given to the students as homework problems. More than half of the students completed the proofs.)

Theorem 13.2. A covering $(D';O') \xrightarrow{f} (D;O)$ is a Galois covering if and only if $f_*(\pi_1(D';O'))$ is a normal subgroup of $\pi_1(D;O)$.

The proof is left to the reader.

Problem: Let $(D';O') \xrightarrow{f'} (D;O)$ and $(D'';O'') \xrightarrow{f''} (D;O)$ be coverings. Let $\Gamma' = f'_*(\pi_1(D';O'))$ and $\Gamma'' = f''_*(\pi_1(D'';O''))$. Prove that $\Gamma' \subset \Gamma''$ if and only if there is a covering $(D';O') \xrightarrow{g} (D'';O'')$ such that $f'' \circ g = f'$.

If such g exists, we say that $(D';O') \xrightarrow{f'} (D;O)$ is higher than $(D'';O'') \xrightarrow{f''}$ $(D;O)$, and that $(D'';O'') \xrightarrow{f''} (D;O)$ is lower than $(D';O') \xrightarrow{f'} (D;O)$.

A class of coverings of D under the equivalence relation defined before Theorem 13.1 is called a *covering class*.

Let us summarize what we have learned so far.

Integrated Theorem 13.3.

A covering $(D'; O') \xrightarrow{f} (D; O)$ over $(D; O)$ induces an injective homomorphism f_* from $\pi_1(D'; O')$ to $\pi(D; O)$:

$$f_* : \pi_1(D'; O') \longrightarrow \pi_1(D; O).$$

Let $\Gamma' = f_*(\pi_1(D'; O'))$, which is a subgroup of $\pi_1(D; O)$. Then Γ' is a normal subgroup if and only if $(D'; O') \xrightarrow{f} (D; O)$ is a Galois covering. If so, then $\pi_1(D'; O')/\Gamma' \cong \Gamma(D' \xrightarrow{f} D)$.

Also, there is a one-to-one correspondence between the set of all covering classes and the set of all subgroups of $\pi_1(D; O)$ given by $[(D'; O') \xrightarrow{f} (D; O)] \longleftrightarrow \Gamma' = f_*(\pi_1(D'; O'))$. In particular, the universal covering space corresponds to $\Gamma' = \{1\}$, and the high-low relation between coverings corresponds to the inclusion relation of subgroups.

Seeing Galois Theory

The Fourteenth Week:
Continuous functions on covering surfaces

Starting this week, we will look at functions defined on manifolds. Let D be a two-dimensional manifold. (If you still feel uncomfortable hearing the word "manifold", think of D as a region in a plane.) We will denote by $\mathbf{C}$ the set of all complex numbers as usual, and consider a continuous function $F : D \longrightarrow \mathbf{C}$ that assigns a complex number $F(P)$ to each point $P \in D$. The symbol $C^0(D)$ stands for the set of all continuous functions on D.

It is easy to see that if F and G are continuous on D, the sum, difference, and product of F and G are also continuous. Here, we define the sum, difference, and product as

$$F + G : D \ni P \mapsto F(P) + G(P)$$
$$F - G : D \ni P \mapsto F(P) - G(P)$$
$$F \cdot G : D \ni P \mapsto F(P) \cdot G(P)$$

respectively. (Thus $C^0(D)$ is a commutative ring.)

The quotient F/G cannot always be defined. If G is never 0 on D, then

$$F/G : D \ni P \mapsto F(P)/G(P)$$

is defined and continuous.

We may identify constant functions

$$c : D \ni P \mapsto c \in \mathbf{C}$$

with complex numbers $c \in \mathbf{C}$. Therefore, we can say that $C^0(D)$ contains $\mathbf{C}$ as a subring.

Let $D' \xrightarrow{f} D$ be a covering of D. If F is a continuous function on D, then the composition $F \circ f$ is also continuous (on D'), because the composition of two continuous functions is continuous. This produces a map from $C^0(D)$ to $C^0(D')$ defined by

$$C^0(D) \ni F \mapsto F \circ f \in C^0(D').$$

We denote this map by f^*:

$$f^* : C^0(D) \longrightarrow C^0(D')$$

$$f^*(F) = F \circ f$$

The following theorem is easy to see.

Theorem 14.1.

(a) $(F \pm G) \circ f = F \circ f \pm G \circ f$; i.e., $f^*(F + G) = f^*(F) + f^*(G)$.

(b) $(F \cdot G) \circ f = (F \circ f) \cdot (G \circ f)$; i.e., $f^*(F \cdot G) = f^*(F) \cdot f^*(G)$.

(c) If $F \neq G$, then $F \circ f \neq G \circ f$; i.e., $f^*(F) \neq f^*(G)$.

This means that f^* is an injective ring homomorphism from $C^0(D)$ to $C^0(D')$. Therefore, $C^0(D)$ is ring isomorphic to the subring $f^*(C^0(D))$ of $C^0(D')$: $C^0(D) \cong f^*(C^0(D)) \subset C^0(D')$.

Proof: (a)

$$
\begin{aligned}
(F + G) \circ f(P) &= (F + G)(f(P)) \\
&= F(f(P)) + G(f(P)) \\
&= (F \circ f)(P) + (G \circ f)(P) \\
&= (F \circ f + G \circ f)(P) \text{ for all } P \in D'.
\end{aligned}
$$

Thus, $(F + G) \circ f = F \circ f + G \circ f$.

(b) Similar.

(c) If $F \neq G$, then there is a point $P_0 \in D$ such that $F(P_0) \neq G(P_0)$. Since f is surjective, there is a point $P_0' \in D'$ with $f(P_0') = P_0$. Then $(F \circ f)(P_0') = F(P_0) \neq G(P_0) = (G \circ f)(P_0')$. Thus, $f^*(F) \neq f^*(G)$.   Q.E.D.

If we identify F and $f^*(F) = F \circ f$, then $C^0(D)$ becomes a subring of $C^0(D')$. We will make this identification starting in the 18th week.

Preparation Theorem. A function $F' \in C^0(D')$ belongs to $f^*(C^0(D))$ if and only if F' has the same value for conjugate points of the covering $D' \xrightarrow{f} D$.

Proof: Two points P' and Q' are conjugate if $f(P') = f(Q')$, by definition. If $F' \in f^*(C^0(D))$, then $F' = f^*(F) = F \circ f$ for some $F \in C^0(D)$. Therefore, $F'(P') = F(f(P')) = F(f(Q')) = F'(Q')$. In other words, F' takes the same value on conjugate points.

Conversely, let F' be a continuous function on D' that has the same value on points which are conjugate. If we set $f^{-1}(P) = \{P_1', P_2', \ldots\}$ for $P \in D$, then $P_1', P_2', \ldots$ are all conjugate. Therefore, $F'(P_1') = F'(P_2') = F'(P_3') = \ldots$. Since the common value $F'(P_1') = F'(P_2') = \ldots$ depends only on P, we can denote this value by $F(P)$. This F is a function on D, and it is easy to show that F is continuous.

In order to see this, it suffices to show that F is continuous on an arbitrary neighborhood in D, since continuity is a local property. Take a copiable neighborhood U around P and its copy U_1' containing P_1'. Choose a point Q in U and its shadow Q_1' in U_1'. Then $F(Q) = F(Q_1')$ by the definition of F. Since $Q_1' = (f|_{U_1'})^{-1}(Q)$, and $f|_{U_1'}$ is a homeomorphism between U_1' and U, $F = F' \circ (F|_{U_1'})^{-1}$ is a continuous function. (End of proof of continuity.) Now if we define $f^*(F) = F \circ f$ for the continuous function F, then this is nothing but F'. (End of proof of Preparation Theorem.)

In what follows, we consider Galois coverings $f : D' \longrightarrow D$. We denote by Γ the covering transformation group $\Gamma(D' \overset{f}{\to} D)$. Let γ be an element of Γ. Recall that γ is a homeomorphism from D' onto itself satisfying $f \circ \gamma = f$. For a continuous function F' in $C^0(D')$, $F' \circ \gamma$ is also a continuous function on D'. This defines a map

$$C^0(D') \ni F' \mapsto F' \circ \gamma \in C^0(D')$$

which we denote by γ^*:

$$\gamma^* : C^0(D') \longrightarrow C^0(D')$$
$$\gamma^*(F') = F' \circ \gamma$$

In fact, γ^* is a ring endomorphism of $C^0(D')$. In other words, $\gamma^*(F' \pm G') = \gamma^*(F') \pm (G')$ and $\gamma^*(F' \cdot G') = \gamma^*(F') \cdot \gamma^*(G')$. The proof is similar to that of Theorem 14.1.

For any two elements γ_1 and γ_2 of Γ, consider γ_1^* and γ_2^*. Then $(\gamma_1 \circ \gamma_2)^* = \gamma_2^* \circ \gamma_1^*$. Indeed, for any $F' \in C^0(D')$, $(\gamma_1 \circ \gamma_2)^*(F') = F' \circ (\gamma_1 \circ \gamma_2) = (F' \circ \gamma_1) \circ \gamma_2 = \gamma_2^*(F' \circ \gamma_1) = (\gamma_1^* \circ \gamma_2^*)(F')$. This means that the correspondence $\Gamma \ni \gamma \mapsto \gamma^*$ is an anti-homomorphism from Γ to the group of automorphisms $\mathrm{Aut}(C^0(D'))$ of the ring $C^0(D')$.

Problem: Show that $\gamma \mapsto \gamma^*$ is an anti-isomorphism from Γ to $\mathrm{Aut}(C^0(D'))$. The only thing left to show is that $\gamma_1^* \neq \gamma_2^*$ if $\gamma_1 \neq \gamma_2$.

A function F' of $C^0(D')$ is said to be Γ-*invariant* if $\gamma^*(F') = F'$ for any element γ of Γ. The set of all Γ-invariant functions will be denoted by $C^0(D')^\Gamma$.

$$C^0(D')^\Gamma = \{F' \in C^0(D') \mid \gamma^*(F') = F' \; \forall \, \gamma \in \Gamma\}.$$

Theorem 14.2. $f^*(C^0(D)) = C^0(D')^\Gamma$.

This means that a function of the form $f^*(F) = F \circ f$ is Γ-invariant, and these are the only Γ-invariant functions.

Proof: If $F' = f^*(F)$, F' is clearly Γ-invariant: $\gamma^*(F') = F' \circ \gamma = F \circ f \circ \gamma = F \circ f = F'$. Conversely, let F' be a Γ-invariant function, and P_1' and P_2' a pair of conjugate points in D'. Since $D' \xrightarrow{f} D$ is a Galois covering, there is exactly one $\gamma \in \Gamma$ such that $\gamma(P_1') = P_2'$ by Theorem 11.1. (There, we used the notation $\gamma = \sigma(P_1'; P_2')$.) Now,

$$F'(P_2') = F'(\gamma(P_1')) = (F' \circ \gamma)(P_1') = F'(P_1')$$

$$\uparrow (\Gamma\text{-invariance of } F'.)$$

Therefore, F' takes the same value on conjugate points. By the Preparation Theorem, $F' = f^*(F)$ for some $F \in C^0(D)$. Q.E.D.

The Fifteenth Week:
Function Theory on Covering Spaces

This week we begin function theory on D and D'. As usual, $\mathbf{C}$ denotes the complex plane. For simplicity, we will assume that D is a region (*i.e.*, open, connected subset) in $\mathbf{C}$, and D' is a covering surface of D. This restriction is necessary to rapidly develop function theory on these surfaces.*

Points of our region D will be denoted by $z, z', \ldots$ so that they look like complex numbers. We assume you know the basics of complex function theory. To make sure of this, here is a brief review:

A function F defined on D is *holomorphic* if $F(z)$ is differentiable and $F'(z)$ is continuous. Here F is defined to be differentiable if the limit

$$F'(z) = \lim_{w \to z} \frac{F(w) - F(z)}{w - z}$$

exists. Recall that the limit exists if and only if $\dfrac{F(w) - F(z)}{w - z}$ approaches a finite value no matter how w approaches z. Thus differentiability is a stronger condition in this situation than for real variables. Differentiability of a complex function F is equivalent to the condition that F satisfy the *Cauchy-Riemann equations*:

$$\frac{\partial u}{\partial x} = \frac{\partial v}{\partial y}, \quad \frac{\partial u}{\partial y} = -\frac{\partial v}{\partial x}, \text{ where } F = u + iv.$$

Let $O(D)$ be the set of all holomorphic functions on D. Then $O(D)$ is a ring. That is, if $F, G \in O(D)$, then $F \pm G$ and $F \cdot G \in O(D)$. Recall, also, that if F is holomorphic, then F' is also holomorphic. Therefore, $F''(z), F'''(z), \ldots$ are all holomorphic, *i.e.*,

$$F \in O(D) \implies F', F'', F''', \ldots \in O(D).$$

For a holomorphic function F on D and a curve C in D, we can define the definite integral $\int_C F(z)dz$. To do this, take $n + 1$ points $z_0, z_1, z_2, \ldots, z_n$ on C

---

* Here I wondered whether I should explain complex structures on manifolds, but I abandoned the idea and settled on this presentation. However, a few smart-aleck students didn't like my decision.

starting from the initial point so that $z_0 =$ initial point and $z_n =$ terminal point. Take a representative point ζ_i on a small arc $\overline{z_{i-1}z_i}$ of C and make an approximating sum

$$\sum_{i=1}^{n} F(\zeta_i)(z_i - z_{i-1}).$$

By increasing n (making the subdivision of C smaller), we define

$$\int_C F(z)dz = \lim \sum_{i=1}^{n} F(\zeta_i)(z_i - z_{i-1}).$$

(limit as subdivision is made finer)

When C is a closed curve in D and is null-homotopic, then $\int_C F(z)dz = 0$ (Cauchy's Theorem). This property, in turn, characterizes holomorphicity of F (Morera's Theorem).

By Cauchy's theorem, we can see that an antiderivative exists for $F \in O(D)$ when D is simply connected: Let $a \in D$ and choose a curve C connecting a to $z_0 \in D$. Now $\int_C F(z)dz$ does not depend on the choice of C because D is simply-connected. Therefore, we can write the integral as $\int_C F(z)dz = \int_a^{z_0} F(z)dz$. Now, by varying z_0, we define a function on D which turns out to be an antiderivative of $F(z)$.

Remark: When D is not simply connected, $F(z)$ may not have an antiderivative defined on D. This is because $\int_C Fdz$ could depend on the homotopy class of the curve C connecting a and z_0. In this case, we can construct an antiderivative of F considered as a function on the universal cover $\tilde{D}$ of D. (We will prove this later.)

A holomorphic function F can be expressed as a convergent power series

$$F(z) = \sum_{n=0}^{\infty} \alpha_n(z - a)^n = \alpha_0 + \alpha_1(z - a) + \alpha_2(z - a)^2 + \cdots$$

in a neighborhood of a. In particular, this means that the set of zeros of a holomorphic function F on D does not have an accumulation point in D unless $F \equiv 0$. Therefore $F(z)$ has at most a countable number of zeros. This leads us to the following important corollary: If the product $F(z) \cdot G(z) = 0$ for all $z \in D$, then $F(z) \equiv 0$ or $G(z) \equiv 0$. That is, the ring $O(D)$ has no zero divisors. (Hence it's an integral domain.) So we can form the quotient field of $O(D)$. An element of this quotient field is a meromorphic function on D. A function is *meromorphic* if it is holomorphic except at a countable number of points (*poles*) which do not accumulate in D. Let us denote the set of all meromorphic functions on D by $K(D)$. By construction, $K(D)$ is a field.

Next, let us develop the theory of functions on a covering surface. Let D be a region in $\mathbf{C}$ and $D' \xrightarrow{z} D$ a covering. (When $D \subset \mathbf{C}$, it will be convenient later to use the letter z instead of f to denote the covering map.) For a point $P' \in D'$, the value $z(P')$ is a point of D; hence it is a complex number. The map z is, by definition, a continuous function: $z \in C^0(D')$. When F and G are continuous on D', we say that F is *differentiable at P' with respect to G* if the limit

$$\lim_{Q' \to P'} \frac{F(Q') - F(P')}{G(Q') - G(P')}$$

is defined and finite. Here, the limit should exist independently of the way Q' approaches P'. We denote the limit by $\dfrac{\partial F}{\partial G}(P')$.

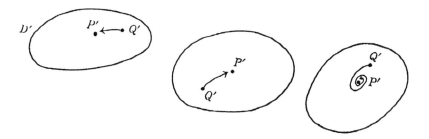

Figure 15.1. Examples of different ways of approaching P'

[Aside]: A simplistic explanation like this probably confuses the bright students.

Now, the precise definition:

Suppose that there is a neighborhood U' of P' in D' such that $G(Q') \neq G(P')$ for $Q' \in U' - \{P'\}$. Suppose also that $\dfrac{F(Q') - F(P')}{G(Q') - G(P')}$, considered as a function of Q' on $U' - \{P'\}$, can be extended to a continuous function, which we call $\dfrac{\partial F}{\partial G}$, on U'. Then we say that $\dfrac{\partial F}{\partial G}(P') = \lim\limits_{Q' \to P'} \dfrac{F(Q') - F(P')}{G(Q') - G(P')}$ exists.

In particular, if a function $F \in C^0(D')$ is differentiable at P' with respect to the function z, we will simply say that F is differentiable at P'. If a function $F \in C^0(D')$ is differentiable (with respect to z) at each point of D', we have the derivative

$$\frac{\partial F}{\partial z} : D' \ni P' \mapsto \frac{\partial F}{\partial z}(P').$$

If, further, $\dfrac{\partial F}{\partial z}$ is continuous on D', we say that F is holomorphic. The set of all holomorphic functions on D' will be denoted by $O(D')$. In particular, z is a holomorphic function, and $\dfrac{\partial z}{\partial z} \equiv 1$.

In the same way, we can define holomorphic functions on an open subset U' of D'. The set of all holomorphic functions on U' will be denoted by $O(U')$.

Since the property of a function being holomorphic is a local one, we see the following immediately: Suppose D' is covered by at most a countable number of open neighborhoods U_i' ($i = 1, 2, 3, \ldots$), *i.e.*, $U' = \bigcup_{i=1}^{\infty} U_i'$. A function $F \in C^0(D')$ is holomorphic if and only if each restriction $F|_{U_i'}$ ($i = 1, 2, \ldots$) is holomorphic on U_i'.

As a special case, consider when U' is small enough so that it is a copy of a copiable neighborhood U of D. Then the restriction $z|_{U'}$ of the covering map z to U' is a homeomorphism. Hence the inverse $(z|_{U'})^{-1}$ exists. The fact that $F \in O(U')$ is equivalent to saying that $F \circ (z|_{U'})^{-1}$ is holomorphic (in the usual sense) on U.

Therefore, we could define F to be holomorphic when the following condition is satisfied: For an arbitrary copiable neighborhood U in D and a copy U' of U in D', the function $F \circ (z|_{U'})^{-1}$ defined on U is holomorphic (in the ordinary sense of function theory).

We can prove theorems concerning these "holomorphic" functions similar to those in the ordinary theory of functions.

First, if F and G are holomorphic in D', so are $F \pm G$ and $F \cdot G$. Therefore, $O(D')$ is a commutative ring. Also, $F \in O(D')$ implies that $\dfrac{\partial F}{\partial z}, \dfrac{\partial^2 F}{\partial z^2}, \ldots \in O(D')$. We define the integral $\int_C F dG$ with respect to $F, G \in O(D')$ and a curve C in D' as follows: Take $n + 1$ points $P_0, P_1, \ldots, P_n$ on C so that $P_0 = $ initial point and $P_n = $ terminal point. Take a representative point Q_i in each small arc $\overline{P_{i-1} P_i}$ and make an approximating sum

$$\sum_{i=1}^{n} F(Q_i)\{G(P_i) - G(P_{i-1})\}.$$

The limit of this sum as the subdivision of C becomes finer (that is, each arc gets as small as you wish) is denoted by $\int_C F dG$:

$$\int_C F dG = \lim \sum_i F(Q_i)\{G(P_i) - G(P_{i-1})\}$$

(limit as subdivision is made finer)

The existence of this limit can be shown in the same way that ordinary line integrals are defined, so we omit the proof here.

In particular, integrals of the form $\int_C F dz$ are important, because $\int_C F dG = \int_C (F \frac{dG}{dz}) dz$. If C is null-homotopic in D' and $F \in O(D')$, then (by Cauchy)

$$\int_C F(z) dz = 0.$$

Conversely, this property characterizes the holomorphicity of F (Morera). The proofs are similar to those in the usual theory of functions. Cauchy's Theorem implies that when D' is simply connected, $\int_{Q'}^{P'} F dz$ is uniquely determined by $F \in O(D')$ and two points $P', Q' \in D'$, and F has an antiderivative.

A holomorphic function F on D' can be expanded as a convergent power series locally: if A is a point in D' and P' another point in D' close to A, then we can write

$$F(P') = \sum_{n=0}^{\infty} \alpha_n (z(P') - z(A))^n.$$

This implies that the set of zeros of a holomorphic function F (with $F \not\equiv 0$) has no accumulation points in D'. Therefore this set is discrete in D'. This, in turn, implies that $O(D')$ is an integral domain. The quotient field of $O(D')$ is the set of meromorphic functions on D'. To paraphrase this, a function ψ is meromorphic on D' if

(i) ψ is defined and holomorphic on $D' - \{A_1, A_2, \ldots\}$, where $\{A_1, A_2, \ldots\}$ is a discrete subset of D'.

(ii) In a neighborhood of A_i, we can expand ψ as

$$\psi(P') = \sum_{n=-d_i}^{\infty} \alpha_n (z(P') - z(A_i))^n$$
$$= \frac{\alpha_{-d_i}}{(z(P') - z(A_i))^{d_i}} + \frac{\alpha_{-d_i+1}}{(z(P') - z(A_i))^{d_i-1}} + \cdots$$
$$+ \alpha_0 + \alpha_1(z(P') - z(A_i)) + \alpha_2(z(P') - z(A_i))^2 + \cdots$$

(In other words, ψ has a Laurent series expansion with finite principal part.) The set of meromorphic functions on D' is denoted by $K(D')$, so $K(D')$ is a field.

Theorem 15.1. Let D be a region in $\mathbf{C}$, and $D' \overset{z}{\to} D$ a covering of D. We have $z^*(O(D)) \subset O(D')$, where $z^*(F) = F \circ z$. Further,

$$O(D') \cap z^*(C^0(D)) = z^*(O(D)).$$

For any $F \in O(D)$,

$$z^*\left(\frac{dF}{dz}\right) = \frac{dz^*(F)}{dz}$$

Proof: Let P' and Q' be two points of D' and $z(P') = z$, $z(Q') = z_0$. For $F \in C^0(D)$,

$$\frac{z^*(F)(P') - z^*(F)(Q')}{z(P') - z(Q')} = \frac{F(z(P')) - F(z(Q'))}{z(P') - z(Q')}$$
$$= \frac{F(z) - F(z_0)}{z - z_0} \tag{$*$}$$

This is a trivial formula. But since $P' \longrightarrow Q'$ implies $z \longrightarrow z_0$,

$$\lim_{P' \to Q'} \frac{z^*(F)(P') - z^*(F)(Q')}{z(P') - z(Q')} = \lim_{z \to z_0} \frac{F(z) - F(z_0)}{z - z_0} = \frac{dF}{dz}(z_0).$$

In other words, if F is differentiable at z_0, then $z^*(F)$ is differentiable at Q', and the derivative

$$\frac{d(z^*(F))}{dz}(Q') = \frac{dF}{dz}(z_0) = \frac{dF}{dz}(z(Q')).$$

Therefore, if F is holomorphic on D, then $z^*(F)$ is holomorphic on D', and the derivatives satisfy the relation

$$\frac{dz^*(F)}{dz}(P') = \frac{dF}{dz}(z(P')) = z^*\left(\frac{dF}{dz}\right)(P').$$

This is equivalent to saying that

$$\frac{d(z^*F)}{dz} = z^*\left(\frac{dF}{dz}\right)$$

Conversely, in equation $(*)$, let U be a copiable neighborhood, U' be its copy containing Q', $z \in U$, and $P' \in U'$. If $z \longrightarrow z_0$, then $P' \longrightarrow Q'$. Therefore, if $z^*(F)$ is differentiable at Q', then

$$\lim_{z \to z_0} \frac{F(z) - F(z_0)}{z - z_0} = \lim_{P' \to Q'} \frac{z^*(F)(P') - z^*(F)(Q')}{z(P') - z(Q')} = \frac{dz^*(F)}{dz}(Q')$$

exists. Hence F is differentiable at z_0, and its derivative is $\dfrac{dz^*(F)}{dz}(Q')$. Therefore, if $z^*(F)$ is holomorphic, F is also holomorphic, because $\dfrac{dz^*(F)}{dz}$ is continuous. This means that $z^*(C^0(D)) \cap O(D') = z^*(O(D))$. Q.E.D.

Theorem 15.2. Let $D \subset \mathbf{C}$ and $D' \xrightarrow{z} D$ be a Galois covering. Denote the covering transformation group $\Gamma(D' \xrightarrow{z} D)$ by Γ. Then, for $F' \in O(D')$ and $\gamma \in \Gamma$, $\gamma^*(F')$ is also holomorphic. Further, we have

$$\frac{d\gamma^*(F')}{dz} = \gamma^*\left(\frac{dF'}{dz}\right) \qquad (\forall \gamma \in \Gamma).$$

If we define $O(D')^\Gamma = O(D') \cap C^0(D')^\Gamma$, then

$$z^*(O(D)) = O(D')^\Gamma.$$

Proof: From the definition of covering transformations, we know that $z = z \circ \gamma$ if $\gamma \in \Gamma$. For two points $P', Q' \in D'$, denote $\gamma(P') = R'$, and $\gamma(Q') = S'$. Then

$$\frac{\gamma^*(F')(P') - \gamma^*(F')(Q')}{z(P') - z(Q')} = \frac{F'(\gamma(P')) - F'(\gamma(Q'))}{z \circ \gamma(P') - z \circ \gamma(Q')} = \frac{F'(R') - F'(S')}{z(R') - z(S')}$$

Since $P' \longrightarrow Q'$ implies that $R' \longrightarrow S'$,

$$\lim_{P' \to Q'} \frac{\gamma^*(F')(P') - \gamma^*(F')(Q')}{z(P') - z(Q')} = \lim_{R' \to S'} \frac{F'(R') - F'(S')}{z(R') - z(S')}.$$

Therefore, $\gamma^*(F')$ is also holomorphic, and

$$\frac{d\gamma^*(F')}{dz}(Q') = \frac{dF'}{dz}(S') = \frac{dF'}{dz}(\gamma(Q')) = \gamma^*\left(\frac{dF'}{dz}\right)(Q').$$

In other words,

$$\frac{d\gamma^*(F')}{dz} = \gamma^* \frac{dF'}{dz}.$$

Further, by using Theorem 14.1 and Theorem 15.1, we have

$$z^*(O(D)) = O(D') \cap z^*(C^\circ(D)) = O(D') \cap C^0(D')^\Gamma = O(D')^\Gamma.$$

Q.E.D.

The first half of Theorem 15.2 shows that γ induces a ring automorphism γ^* of $O(D')$. That is,

(1) $\gamma^*(O(D')) = O(D')$

(2) $\gamma^*(F \pm G) = \gamma^*(F') \pm \gamma^*(G')$ and $\gamma^*(F' \cdot G') = \gamma^*(F') \cdot \gamma^*(G')$.

((2) was proved last week.)

Furthermore, γ^* induces an automorphism of the field $K(D')$ of meromorphic functions on D'. When a meromorphic function $\psi \in K(D')$ has poles $\{A'_1, A'_2, A'_3, \ldots\}$, and is holomorphic on $D'_0 = D' - \{A'_1, A'_2, A'_3, \ldots\}$, the function $\psi \circ (\gamma|_{\gamma^{-1}(D'_0)})$ defined on $\gamma^{-1}(D'_0) = D' - \{\gamma^{-1}(A'_1), \gamma^{-1}(A'_2), \ldots\}$ naturally becomes a meromorphic function of D' with $\{\gamma^{-1}(A'_i)\}$ as the set of poles. We denote this meromorphic function by $\gamma^*(\psi) = \psi \circ (\gamma|_{\gamma^{-1}(D'_0)})$. Sometimes we use $\psi \circ \gamma$ if there is no chance of confusion.

Actually, a meromorphic function can be regarded as a holomorphic map from D' to the Riemann sphere $\mathbf{C} \cup \{\infty\}$. In this sense, $\psi \circ \gamma$ is just the composition of two maps.

It is easy to see that the correspondence we obtained by $\psi \mapsto \gamma^*(\psi)$ $(\psi \in K(D'))$ is an automorphism of $K(D')$. In other words, the following statements (3), (4), and (5) are true.

(3) $\gamma^*(K(D')) = K(D')$.

(4) $\gamma^*(\varphi \pm \psi) = \gamma^*(\varphi) \pm \gamma^*(\psi)$, $\gamma^*(\varphi \cdot \psi) = \gamma^*(\varphi) \cdot \gamma^*(\psi)$.

(5) If $\psi \neq 0$, then $\gamma^*(\varphi/\psi) = \gamma^*(\varphi)/\gamma^*(\psi)$.

The next property is easy to check.

(6) $(\gamma_1 \circ \gamma_2)^* = \gamma_2^* \circ \gamma_1^*$ $(\gamma_1, \gamma_2 \in \Gamma)$.

Hence the correspondence $\gamma \mapsto \gamma^*$ is an anti-homomorphism from the group Γ to the group of automorphisms of $K(D')$.

Theorem 15.3. Let $D' \overset{z}{\to} D \subset \mathbf{C}$ be a Galois covering, and $\Gamma = \Gamma(D' \overset{z}{\to} D)$. Then we have

$$z^*(K(D)) = K(D')^{\Gamma}.$$

Proof: The inclusion $z^*(K(D)) \subset K(D')^{\Gamma}$ is obvious. Let's prove the inclusion in the other direction. Take ψ in $K(D')^{\Gamma}$. Denote the poles of ψ by $A_1', A_2', \ldots$. If γ is an arbitrary element of Γ, $\gamma^*(\psi)$ has $\{\gamma^{-1}(A_1'), \gamma^{-1}(A_2'), \ldots\}$ as the set of poles. But since $\gamma^*(\psi) = \psi$ by hypothesis, $\{A_1', A_2', \ldots\} = \{\gamma^{-1}(A_1'), \gamma^{-1}(A_2'), \ldots\}$ as sets. This means that if A_i' is a pole, then any point conjugate to A_i' is also a pole. Therefore, if we take the distinct elements $\{a_1, a_2, a_3, \ldots\}$ of $\{z(A_i') \mid i = 1, 2, 3, \ldots\}$, then $\{A_i' \mid i = 1, 2, 3, \ldots\} = z^{-1}(a_1) \cup z^{-1}(a_2) \cup z^{-1}(a_3) \cup \ldots$. Denote $D - \{a_1, a_2, a_3, \ldots\} = E$, and $D' - \{A_1', A_2', \ldots\} = E'$. If we also denote the restriction of z to E' by z, then z maps E' onto E. It is obvious that $E' \overset{z}{\to} E$ is a covering. Further, it is a Galois covering which corresponds to the same covering transformation group Γ:

$$
\begin{array}{ccc}
E' & \subset & D' \\
\downarrow & \Gamma & \downarrow \\
E & \subset & D
\end{array}
$$

Since ψ can be considered as a holomorphic function on E' ($\psi \in O(E')$), Theorem 15.2 implies that $\psi = z^*(\phi)$ for some $\phi \in O(E)$. Noticing that ψ can be expanded as a Laurent series with finite principal part, we conclude that ϕ must be meromorphic on D as well. Thus, $\psi = z^*\phi = \phi \circ z$, $\phi \in K(D)$.   Q.E.D.

Theorem 15.4. If $D' \overset{z}{\to} D \subset \mathbf{C}$ is a Galois covering, and $\psi \in K(D')$, we have

$$\gamma^*\left(\frac{d\psi}{dz}\right) = \frac{d(\gamma^*\psi)}{dz} \quad \forall \gamma \in \Gamma = \Gamma(D' \overset{z}{\to} D).$$

This is obvious.

Because $z^* : F \mapsto z^*(F)$ is an injective ring homomorphism, we can identify $z^*(F)$ and F. Then $C^0(D) \subset C^0(D')$, $O(D) \subset O(D')$, $K(D) \subset K(D')$. Further,

$\frac{d}{dz}(z^*(F)) = z^*\left(\frac{dF}{dz}\right)$ implies that this identification is compatible with the differ-

ential operator $\frac{d}{dz}$. In other words, the results of differentiating F as a function of D and as a function of D' are the same (again by the above identification).

With these identifications, we can rewrite some of the previous results as

---

Theorem 15.5.

(1) $O(D) = O(D') \cap C^0(D)$.

(2) if $D' \xrightarrow{z} D$ is a Galois covering, and $\Gamma = \Gamma(D \xrightarrow{z} D)$, then

$$O(D) = O(D')^\Gamma$$
$$K(D) = K(D')^\Gamma$$

We will use this identification for the following weeks.

Solvable or not?

The Sixteenth Week:
Differential Equations

If D is a simply connected region in $\mathbf{C}$, we have the following existence theorem for solutions of linear ordinary differential equations:

Theorem 16.1. Let $D \subset \mathbf{C}$ be simply connected, and P and Q holomorphic on D. A homogeneous linear differential equation

$$(\#) \qquad\qquad \frac{d^2 w}{dz^2} + P(z)\frac{dw}{dz} + Q(z)w = 0$$

with initial conditions at $z = z_0 \in D$ given by

$$(*) \qquad\qquad \begin{aligned} w(z_0) &= \alpha \\ \frac{dw}{dz}(z_0) &= \beta \end{aligned}$$

has a unique holomorphic solution w on D satisfying both $(\#)$ and $(*)$.

The proof can be found in most texts on differential equations, *e.g.*, Birkhoff-Rota.

The solution of $(\#)$ is uniquely determined by the initial conditions $(*)$, so we use $w(z; z_0, \alpha, \beta)$ to denote the solution.

Let's forget about $(*)$ for a while. We denote by $V_\#$ the set of functions that satisfy the differential equation $(\#)$. Since $(\#)$ is a homogeneous equation, the set $V_\#$ of solutions is a vector space over $\mathbf{C}$:

$$\left. \begin{aligned} V_\# \ni w_1, w_2, \ldots, w_m \\ \mathbf{C} \ni \lambda_1, \lambda_2, \ldots, \lambda_m \end{aligned} \right\} \implies \sum_{i=1}^m \lambda_i w_i \in V_\#.$$

To see this, note that

$$V_\# \ni w_i \iff \frac{d^2 w_i}{dz^2} + P\frac{dw_i}{dz} + Q w_i = 0.$$

Multiply by λ_i and add. Then

$$\frac{d^2}{dz^2}\left(\sum \lambda_i w_i\right) + P\frac{d}{dz}\left(\sum \lambda_i w_i\right) + Q\left(\sum \lambda_i w_i\right) = 0.$$

Thus, $\sum \lambda_i w_i$ is also a solution of (#). Since the differential equation (#) has order 2, the dimension of $V_\#$ over **C** is 2: $\dim_{\mathbf{C}} V_\# = 2$. This can be seen as follows: Define a map

$$\psi : V_\# \ni w \mapsto \psi(w) = \begin{pmatrix} w(z_0) \\ \dfrac{dw}{dz}(z_0) \end{pmatrix} \in \mathbf{C}^2$$

by assigning a two-dimensional vector $\begin{pmatrix} w(z_0) \\ \frac{dw}{dz}(z_0) \end{pmatrix}$ to $w \in V_\#$. (The base point z_0 is kept fixed throughout.) Clearly ψ is a linear map from $V_\#$ to $\mathbf{C}^2$. By the existence theorem above, if $\begin{pmatrix} \alpha \\ \beta \end{pmatrix} \in \mathbf{C}^2$ is an arbitrary vector, there is a unique solution $w \in V_\#$ such that $\begin{pmatrix} w(z_0) \\ \frac{dw}{dz}(z_0) \end{pmatrix} = \begin{pmatrix} \alpha \\ \beta \end{pmatrix}$. Therefore, ψ is bijective. Thus, $V_\# \cong \mathbf{C}^2$ and $\dim_{\mathbf{C}} V_\# = 2$. $\hfill$ Q.E.D.

Now let us consider differential equations on covering surfaces. Let D be a region in **C** which is not necessarily simply connected. Let the universal covering of D be denoted by $\tilde{D} \overset{z}{\to} D$. The points on $\tilde{D}$ will be denoted by small letters $\tilde{p}, \tilde{p}_0$, etc., from now on. Because $\tilde{D}$ is simply-connected we have the following theorem:

Theorem 16.2. Let $\tilde{D} \overset{z}{\to} D \subset \mathbf{C}$ be the universal covering of D and **P** and **Q** be holomorphic on $\tilde{D}$. Let $\tilde{p}_0$ be a point of $\tilde{D}$, and α and β arbitrary complex numbers. Then there is a unique holomorphic function $w(\tilde{p}; \tilde{p}_0, \alpha, \beta)$ that satisfies both

$$(\tilde{\#}) \qquad \frac{d^2w}{dz^2}(\tilde{p}) + \mathbf{P}(\tilde{p})\frac{dw}{dz}(\tilde{p}) + \mathbf{Q}(\tilde{p})w(\tilde{p}) = 0$$

and the initial conditions

$$(\tilde{*}) \qquad \begin{aligned} w(\tilde{p}_0) &= \alpha \\ \frac{dw}{dz}(\tilde{p}_0) &= \beta \end{aligned}$$

Let $V_{\tilde{\#}}$ denote the set of all solutions of $(\tilde{\#})$, without considering the initial conditions $(\tilde{*})$. Then $V_{\tilde{\#}}$ is a two-dimensional vector space over **C**.

The proof is analogous to that of Theorem 16.1 and is omitted.

So far the coefficient functions **P** and **Q** have been elements of $O(\tilde{D})$. From now on, we will treat the case where **P** and **Q** $\in O(D)$. As was stated earlier, we consider $O(D)$ as a subring of $O(\tilde{D})$ by $O(D) = z^*(O(D)) \subset O(\tilde{D})$. By this identification, we can take **P**, **Q** to be members of $O(D)$. To be more precise, this means that there exist holomorphic functions P and Q on D such that $\mathbf{P}(\tilde{p}) = P(z(\tilde{p}))$ and

$\mathbf{Q}(\tilde{p}) = Q(z(\tilde{p}))$. Therefore, equation $(\tilde{\#})$ becomes

$$(\tilde{\#}) \qquad \frac{d^2w}{dz^2}(\tilde{p}) + P(z(\tilde{p}))\frac{dw}{dz}(\tilde{p}) + Q(z(\tilde{p}))w(\tilde{p}) = 0.$$

We adopt the abbreviation

$$(\#) \qquad \frac{d^2w}{dz^2} + P(z)\frac{dw}{dz} + Q(z)w = 0.$$

Now $(\#)$ looks like a differential equation defined on D, but since D may not be simply connected, its solution w may not be a function on D. In general, it is only a function on $\tilde{D}$. Let the vector space of solutions of $(\tilde{\#}) = (\#)$ be denoted by $V_{\tilde{\#}}$ or $V_{\#}$. The space $V_{\#}$ is a subset of $O(D)$, and it is a two-dimensional vector space over $\mathbf{C}$.

The covering $\tilde{D} \xrightarrow{z} D$ is the universal covering, hence the covering transformation group $\Gamma(\tilde{D} \xrightarrow{z} D)$ is isomorphic to $\pi_1(D; O)$. Let Γ stand for $\Gamma(\tilde{D} \xrightarrow{z} D)$. For any element $\gamma \in \Gamma$, we have $\gamma^* : K(\tilde{D}) \longrightarrow K(\tilde{D})$, as we discussed in the previous week. Apply γ^* to both sides of $(\tilde{\#})$. Then

$$0 = \gamma^*(0) = \gamma^*\left(\frac{d^2w}{dz^2} + \mathbf{P}\frac{dw}{dz} + \mathbf{Q}w\right)$$
$$= \frac{d^2\gamma^*w}{dz^2} + \gamma^*(\mathbf{P})\frac{d(\gamma^*w)}{dz} + \gamma^*(\mathbf{Q})(\gamma^*w).$$

But we have assumed $\mathbf{P}, \mathbf{Q} \in O(D)$, hence $\gamma^*(\mathbf{P}) = \mathbf{P}$, and $\gamma^*(\mathbf{Q}) = \mathbf{Q}$ by Theorem 15.2. Therefore, this equation becomes

$$0 = \frac{d^2\gamma^*w}{dz^2} + \mathbf{P}\frac{d(\gamma^*w)}{dz} + \mathbf{Q}(\gamma^*w).$$

That is to say, γ^*w is again a solution of $(\tilde{\#}) = (\#)$. In summary,

---

Theorem 16.3.

$$\left.\begin{array}{l} V_{\#} \ni w \\ \Gamma \ni \gamma \end{array}\right\} \Longrightarrow \gamma^*w \in V_{\#}$$

That is, the restriction $\gamma^*|_{V_{\#}}$ of $\gamma^* : K(\tilde{D}) \longrightarrow K(\tilde{D})$ maps $V_{\#}$ into itself. We denote $\gamma^*|_{V_{\#}}$ by γ^* also. Because we have checked that $\gamma^*(\lambda_1 w_1 + \lambda_2 w_2) = \lambda_1\gamma^*w_1 + \lambda_2\gamma^*w_2$ (cf. Theorem 14.1), $\gamma^* : V_{\#} \longrightarrow V_{\#}$ is a linear map. In particular, if γ is the unit element 1, 1^* is the identity map of $V_{\#}$. The fact $(\gamma_1\gamma_2)^* = \gamma_2^*\gamma_1^*$ implies that $1^* = (\gamma^{-1})^*\gamma^*$ by taking $\gamma_1 = \gamma$, $\gamma_2 = \gamma^{-1}$. Therefore, γ^* has an inverse $(\gamma^*)^{-1} = (\gamma^{-1})^*$. This shows that γ^* is a linear automorphism of $V_{\#}$.

We will adopt the symbol $\mathcal{M}(\gamma)$ to denote $(\gamma^{-1})^*$, which corresponds to the inverse γ^{-1} of γ. From the formula $(\gamma_1\gamma_2)^* = \gamma_2^*\gamma_1^*$, we get $\mathcal{M}(\gamma_1\gamma_2) = \mathcal{M}(\gamma_1)\mathcal{M}(\gamma_2)$. This means that the correspondence

$$\mathcal{M} : \gamma \mapsto \mathcal{M}(\gamma)$$

from the group Γ to the group of linear automorphisms of $V_\#$ is a linear representation of the group Γ.

Let us call this representation $\mathcal{M}$, the *monodromy representation* of (#).

When we fix a basis $[w_1, w_2]$ of $V_\#$, $\mathcal{M}(\gamma)$ can be expressed as a 2×2 matrix. Let the matrix be denoted by

$$M(\gamma) = \begin{pmatrix} a(\gamma) & b(\gamma) \\ c(\gamma) & d(\gamma) \end{pmatrix}.$$

This is, of course, determined by

$$(\mathcal{M}(\gamma)w_1, \mathcal{M}(\gamma)w_2) = ((\gamma^{-1})^*w_1, (\gamma^{-1})^*w_2) = (w_1, w_2)\begin{pmatrix} a(\gamma) & b(\gamma) \\ c(\gamma) & d(\gamma) \end{pmatrix}.$$

The correspondence $M : \Gamma \ni \gamma \mapsto M(\gamma) \in \mathrm{GL}(2, \mathbf{C})$ is a matrix representation of Γ. Here, $\mathrm{GL}(2, \mathbf{C})$ is the group of all nonsingular 2×2 complex matrices. When we change the basis $[w_1, w_2]$, then the representation M changes. However, it is well-known that if T denotes the change of basis matrix, then M changes to $T^{-1}MT$.

Definition: Let $\mathcal{M}$ be the monodromy representation of the differential equation (#). If we can find a matrix representation M corresponding to $\mathcal{M}$ of the form

$$\gamma \mapsto M(\gamma) = \begin{pmatrix} a(\gamma) & b(\gamma) \\ 0 & d(\gamma) \end{pmatrix}$$

by choosing an appropriate basis $[w_1, w_2]$, then we call $\mathcal{M}$ a *triangulable representation*.

The Seventeenth Week:
Elementary methods of solving Differential Equations

Let us consider a set Σ of "known functions". We regard every constant as a known function. The following procedures are used often to produce new functions out of old (known) functions $F_1, F_2, \ldots$:

(i) The four operations of arithmetic:

$$F_1, F_2 \longrightarrow F_1 + F_2$$
$$F_1, F_2 \longrightarrow F_1 - F_2$$
$$F_1, F_2 \longrightarrow F_1 \cdot F_2$$
$$F_1, F_2 \longrightarrow F_1/F_2$$

Linear combinations:

$$F_1, F_2 \longrightarrow \lambda_1 F_1 + \lambda_2 F_2$$

(ii) Differentiation:

$$F \longrightarrow \frac{dF}{dz}$$

(iii) Integration:

$$F(z) \longrightarrow \int F(z)dz$$

(iv) Exponentiation:

$$F(z) \longrightarrow e^{F(z)}$$

We define a process of type L_0 as follows: Starting with $F_1, F_2, \ldots$, apply procedures (i), (ii), (iii), and (iv) finitely many times to produce a new function. For example, the following way of constructing $\dfrac{(F_1 + F_2) \int F_3 dz}{e^{\int F_3 dz}}$ is a process of type L_0:

$$F_1, F_2, F_3 \begin{matrix} \overset{(i)}{\nearrow} & F_1 + F_2 \\ \\ \underset{(iii)}{\searrow} & \int F_3 dz \end{matrix} \Bigg\} \quad \overset{(i)}{\to} \quad (F_1 + F_2) \int F_3 dz \Bigg\} \underset{(iv)}{\searrow} \quad e^{\int F_3 dz} \Bigg\} \quad \overset{(i)}{\to} \quad \frac{(F_1 + F_2) \int F_3 dz}{e^{\int F_3 dz}}$$

The functions obtained by a process of type L_0 starting from known functions (*i.e.*, elements of Σ) will be called *functions of type L_0 on Σ*. This set will be denoted by $L_0(\Sigma)$. For example, when $\Sigma = \{F_1, F_2, F_3, C\}$, where C is a constant,

$$\frac{(F_1 + F_2) \int F_3 dz}{e^{\int F_3 dz}}, \quad e^{-\int F_1 dz} \int F_2 e^{-\int F_1 dz}, \quad \frac{dF_1}{dz} e^{(\int F_2 dz) e^{\int F_3 dz}}$$

are elements of $L_0(\Sigma)$.

We say that a process is of type L if it is a finite sequence of operations of types (i), (ii), (iii), (iv), and (v), where (v) is defined as follows:

(v) Solving algebraic equations:

$$F \longrightarrow \sqrt[n]{F}, \text{ or, more generally,}$$

$$F_1, F_2, \ldots, F_n \longrightarrow \text{ A root } \psi \text{ of } \psi^n + F_1 \psi^{n-1} + F_2 \psi^{n-2} + \cdots + F_n = 0$$

A function obtained from Σ by a process of type L will be called a *function of type L on Σ*. The set of all such functions will be denoted by $L(\Sigma)$. Clearly, $L(\Sigma) \supset L_0(\Sigma)$. For example, if $\Sigma = \{F_1, F_2, F_3, \text{ constant functions}\}$, then

$$\sqrt[7]{e^{-\int F_1 dz} \int F_2 e^{\int F_1 dz}} + \sqrt[5]{F_3} + \sqrt{\frac{dF_2}{dz}} \in L(\Sigma).$$

The letter L in $L_0(\Sigma)$ and $L(\Sigma)$ is used here in honor of Liouville.

Now suppose the coefficient functions $P(z)$ and $Q(z)$ in an equation $\dfrac{d^2 w}{dz^2} + P(z) \dfrac{dw}{dz} + Q(z)w = 0$ belong to the set Σ of known functions. If all the solutions of this differential equation are of type L_0 on Σ, we say that the differential equation is of type L_0 on Σ. If all of the solutions are of type L on Σ, the differential equation is of type L on Σ.

Preparation Theorem 17.1. Suppose that one non-trivial solution of the differential equation $\dfrac{d^2w}{dz^2} + P(z)\dfrac{dw}{dz} + Q(z)w = 0$ is of type L_0 on Σ. Then all the solutions of this equation are of type L_0 on Σ. The statement is also true when L_0 is replaced by L.

Proof: Let the non-trivial solution of type L_0 be denoted by w_1. For an arbitrary function w, define $\dfrac{w}{w_1} = u$, i.e., $w = w_1 u$. Then

$$\frac{dw}{dz} = \frac{dw_1}{dz}u + w_1\frac{du}{dz}$$

$$\frac{d^2w}{dz^2} = \frac{d^2w_1}{dz^2}u + 2\frac{dw_1}{dz}\frac{du}{dz} + w_1\frac{d^2u}{dz^2}.$$

Thus,

$$\frac{d^2w}{dz^2} + P\frac{dw}{dz} + Qw = \left(\frac{d^2w_1}{dz^2} + P\frac{dw_1}{dz} + Qw_1\right)u + 2\frac{dw_1}{dz}\frac{du}{dz} + w_1\frac{d^2u}{dz^2} + Pw_1\frac{du}{dz}$$

$$= w_1\frac{d^2u}{dz^2} + 2\frac{dw_1}{dz}\frac{du}{dz} + Pw_1\frac{du}{dz}$$

(Since w_1 is a solution of the differential equation, the expression inside the parentheses is 0.)

Therefore,

$$w \text{ is a solution} \iff w_1\frac{d^2u}{dz^2} + (2\frac{dw_1}{dz} + Pw_1)\frac{du}{dz} = 0$$

$$\iff \frac{d^2u}{dz^2} + (\frac{2}{w_1}\frac{dw_1}{dz} + P)\frac{du}{dz} = 0$$

Now, this is a differential equation of first order with respect to $\dfrac{du}{dz}$. Thus, we can find a solution by separation of variables. Let $\dfrac{du}{dz} = v$. Then

$$w \text{ is a solution} \iff \frac{1}{v}\frac{dv}{dz} = 2\frac{-1}{w_1}\frac{dw_1}{dz} - P$$

$$\iff \log v = -2\log w_1 - \int P\,dz + C$$

$$\iff \frac{du}{dz} = v = w_1^{-2}e^{-\int P\,dz + C}$$

$$\iff u = \int w_1^{-2}e^{-\int P\,dz + C} + C'$$

$$\iff w = w_1 u = w_1 \cdot \int w_1^{-2}e^{-\int P\,dz + C} + C'$$

All the solutions can be written in this way. Thus, if w_1 is of type L_0 on Σ, so are all the other solutions of this differential equation. The proof is similar for functions of type L.   Q.E.D.

From now on we take $\Sigma = K(D)$. In particular, all the single-valued holomorphic functions on D are "known". However, functions defined on $\tilde{D}$ are considered to be unknown. Now we want to see if the differential equation

$$(\#) = (\tilde{\#}) : \frac{d^2w}{dz^2} + P(z)\frac{dw}{dz} + Q(z)w = 0 \quad P, Q \in O(D)$$

is of type L_0 on $\Sigma = K(D)$.

Theorem 17.2. The equation $(\#) = (\tilde{\#})$ is of type L_0 on $\Sigma = K(D)$ if and only if the monodromy representation $\mathcal{M}$ of $(\#)$ is triangulable.

The "only if" part of the proof is omitted here because it is difficult. The "if" part is easy. Assume that the monodromy representation is triangulable. We will show that $(\tilde{\#})$ is of type L_0 on Σ.

Since $\mathcal{M}$ is triangulable, we can find a basis $[w_1, w_2]$ of $V_\#$ so that for any $\gamma \in \Gamma$,

$$((\gamma^{-1})^*w_1, (\gamma^{-1})^*w_2) = (\mathcal{M}(\gamma)w_1, \mathcal{M}(\gamma)w_2)$$

$$= (w_1, w_2)\begin{pmatrix} a(\gamma) & b(\gamma) \\ 0 & d(\gamma) \end{pmatrix}.$$

This means that

$$w_1 \circ \gamma^{-1} = (\gamma^{-1})^*w_1 = w_1 a(\gamma) \tag{1}$$

$$w_2 \circ \gamma^{-1} = (\gamma^{-1})^*w_2 = w_1 b(\gamma) + w_2 d(\gamma). \tag{2}$$

Differentiate both sides of (1); we get

$$\frac{dw_1}{dz}a(\gamma) = \frac{d((\gamma^{-1})^*w_1)}{dz} = (\gamma^{-1})^*\frac{dw_1}{dz}. \tag{3}$$

Take the quotient of (3) by (1):

$$\frac{\dfrac{dw_1}{dz}}{w_1} = \frac{(\gamma^{-1})^*\dfrac{dw_1}{dz}}{(\gamma^{-1})^*w_1} = (\gamma^{-1})^*\left(\frac{\dfrac{dw_1}{dz}}{w_1}\right). \tag{4}$$

If we put $A = \dfrac{dw_1}{dz}/w_1$, then obviously $A \in K(\tilde{D})$. But (4) says that

$$(\gamma^{-1})^*(A) = A \quad \text{for any element } \gamma \in \Gamma.$$

By Theorem 15.5, we have

$$A \in K(D).$$

Therefore, A is an element of $K(D)$ (a known function.) Integrate both sides of $A = \dfrac{dw_1}{dz}/w_1$, and we obtain

$$\int A dz + C = \log w_1$$

$$\therefore w_1 = e^{\int A dz + C}.$$

Thus, w_1 can be obtained from the known function A by a process of type L_0. This means that $V_{\#}$ contains a function of type L_0 on Σ. Preparation Theorem 17.1 now shows that all elements of $V_{\#}$ are of type L_0. So, by definition, $(\#)$ is of type L_0 on Σ.   Q.E.D.

The Eighteenth Week:
Regular Singularities

First, let us review some basic facts about differential equations with regular singularities. (For details, see Birkhoff-Rota [4], for example.) Consider an open disc $U = U(a; \epsilon)$ of radius ϵ and center a in the complex plane $\mathbf{C}$. Let U_a denote $U - \{a\}$. We call such a region a "5-yen coin." Choose a point b in U_a and let $z : (\tilde{U}_a, \tilde{b}) \longrightarrow (U_a, b)$ be the universal covering. We call $\tilde{U}_a$ a spiral staircase. For simplicity, we assume that $b - a$ is a positive real number.

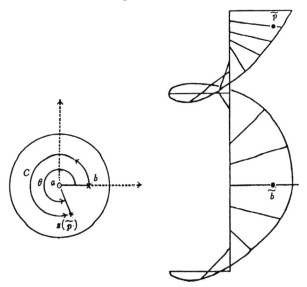

Figure 18.1.

We define the *argument* $\arg(\tilde{p})$ for every point $\tilde{p}$ of $\tilde{U}_a$ as follows. First, let $\tilde{C}$ be a curve in $\tilde{U}_a$ connecting $\tilde{b}$ and $\tilde{p}$, and let C be its projection $z(\tilde{C})$ into U_a. Then C is a curve in U_a starting at b and ending at $z(\tilde{p})$. The homotopy class of C depends only on $\tilde{p}$ and not on the choice of $\tilde{C}$ (because $\tilde{U}_a$ is a universal covering surface). Now $\arg(\tilde{p})$ is defined to be the total angle about the center a swept out by a point which moves along the curve C from b to $z(\tilde{p})$. For a point $\tilde{p}$ in U_a, $\arg\tilde{p}$ has values in $(-\infty, \infty)$. We also define the *modulus* of $\tilde{p}$ as $|z(\tilde{p}) - a|$ and denote it by $r(\tilde{p}) = r(\tilde{p}; a)$. Thus $0 < r(\tilde{p}) < \epsilon$.

A point $\tilde{p}$ of $\tilde{U}_a$ is determined by $\arg(\tilde{p})$ and $r(\tilde{p})$. We call $r(\tilde{p})$ and $\arg(\tilde{p})$ the *polar coordinates* of $\tilde{p} \in \tilde{U}_a$. It is easy to see that $\tilde{p_1}, \tilde{p_2} \in \tilde{U}_a$ are conjugate (that is, $z(\tilde{p_1}) = z(\tilde{p_2})$) if and only if $r(\tilde{p_1}) = r(\tilde{p_2})$ and $\arg(\tilde{p_1}) \equiv \arg(\tilde{p_2}) \pmod{2\pi}$. We define a complex function $\log(z - a)$ on the complex surface $\tilde{U}_a$ in terms of the real function log and the real-valued function arg:

$$\tilde{p} \mapsto \log(r(\tilde{p}; a)) + i\arg(\tilde{p})$$

This is a holomorphic function on $\tilde{U}_a$ and satisfies

$$\frac{d\log(z - a)}{dz}(\tilde{p}) = \frac{1}{z(\tilde{p}) - a}$$

Therefore

$$\int_{\tilde{p_0}}^{\tilde{p}} \frac{dz}{z - a} = (\log(z - a))(\tilde{p}) - (\log(z - a))(\tilde{p_0})$$

For a complex number α, we define a function $(z - a)^\alpha$ by

$$(z - a)^\alpha(\tilde{p}) = e^{\alpha(\log(z-a))(\tilde{p})}$$

Then, obviously, we have

$$|(z - a)^\alpha(\tilde{p})| = r(\tilde{p}; a)^{\mathrm{Re}\,(\alpha)} \cdot e^{-\mathrm{Im}\,(\alpha)\arg(\tilde{p})}$$

The region

$$\{\tilde{p} \in \tilde{U}_a \mid |\arg(\tilde{p})| < K\}$$

in $\tilde{U}_a$ is called a *Stolz sector* of angle $2K$.

Let F be an arbitrary function on $\tilde{U}_a$. If, for every $\epsilon > 0$ and $K > 0$ there exists a positive number $\delta = \delta(\epsilon, K)$ such that $r(\tilde{p}) < \delta$, $|\arg(\tilde{p})| < K$ implies that $|F(\tilde{p}) - c| < \epsilon$, then we write

$$\lim_{\substack{z(\tilde{p}) \to a \\ (Stolz)}} F(\tilde{p}) = c \qquad (*)$$

In polar coordinates (r, θ), we have $F(\tilde{p}) = F(r, \theta)$; then $(*)$ is equivalent to

$$\lim_{r \to 0} F(r, \theta) = c \quad \text{(uniformly, on compact subsets, with respect to } \theta)$$

Formula 18.1.

(a)
$$\lim_{\substack{z(\tilde{p}) \to a \\ (Stolz)}} ((z - a)\log(z - a))(\tilde{p}) = 0.$$

If $\mathrm{Re}\,\alpha > 0$, then

(b)
$$\lim_{\substack{z(\tilde{p}) \to a \\ (Stolz)}} (z - a)^\alpha(\tilde{p}) = 0.$$

Indeed, if $\tilde{p} = (r, \theta)$ (polar coordinates), then $z(\tilde{p}) - a = re^{i\theta}$, and

$$(z - a)\log(z - a)(\tilde{p}) = re^{i\theta}(\log r + i\theta).$$

Thus, $|\theta| < K$ implies

$$\lim_{r \to 0}((z - a)\log(z - a))(\tilde{p}) = \lim_{r \to 0} e^{i\theta}(r \log r + ri\theta) = 0$$

(uniformly, with respect to θ, on compact subsets).

Now let $\alpha = a + bi$. Then $a > 0$ and we have

$$|(z - a)^\alpha(\tilde{p})| = |e^{(a+bi)(\log r + i\theta)}| = e^{a \log r - b\theta} = r^a e^{-b\theta} \to 0 \quad (\text{as } r \to 0)$$

(uniformly, with respect to θ, on compact subsets).
Q.E.D.

Let γ be the class of curves which wind once around a in U_a. Then $\pi_1(U_a; b)$ is the cyclic group $\{\gamma^n \mid n = 0, \pm1, \pm2, \ldots\}$ generated by γ. Since $z : (\tilde{U}_a; \tilde{b}) \to (U_a; b)$ is the universal covering, $\pi_1(U_a; b)$ corresponds to the universal covering group $\Gamma(\tilde{U}_a \xrightarrow{z} U_a)$.

Now let us investigate the action of a covering transformation γ on $\log(z - a)$ and $(z - a)^\alpha$.

Formula 18.2.

(a) $$\gamma^*(\log(z - a)) = \log(z - a) + 2\pi i$$

(b) $$\gamma^*((z - a)^\alpha) = e^{2\pi i\alpha}(z - a)^\alpha$$

Proof: If we set $\tilde{p} = [r, \theta]$, then $\gamma(\tilde{p}) = [r, \theta + 2\pi]$, by the argument in the 6th week. Therefore,

$$(\gamma^* \log(z - a))(\tilde{p}) = \log(z - a)(\gamma(\tilde{p})) = \log r + i(\theta + 2\pi)$$
$$= \log(z - a)(\tilde{p}) + 2\pi i.$$

$$(\gamma^*(z - a)^\alpha)(\tilde{p}) = (z - a)^\alpha(\gamma(\tilde{p})) = e^{\alpha(\log r + i\theta + 2\pi i)}$$
$$= e^{2\pi i\alpha} e^{\alpha(\log r + i\theta)} = e^{2\pi i\alpha}(z - a)^\alpha(\tilde{p}).$$

Q.E.D.

We say that a is a regular singular point of F if the function F on $\tilde{U}_a$ has an expansion of the following type:

$$F(\tilde{p}) = \sum_{i=1}^{m_0} (z-a)^{\alpha_i} A_i(z-a)$$

$$+ \log(z-a) \sum_{i=1}^{m_1} (z-a)^{\beta_i} B_i(z-a)$$

$$+ (\log(z-a))^2 \sum_{i=1}^{m_2} (z-a)^{\gamma_i} C_i(z-a)$$

$$+ \cdots$$

$$+ (\log(z-a))^\nu \sum_{i=1}^{m_\nu} (z-a)^{\omega_i} W_i(z-a)$$

where $A_i(t), B_i(t), C_i(t), \ldots, W_i(t)$ are convergent power series on $|t| < \epsilon$ and $\alpha_i, \beta_i, \gamma_i, \ldots, \omega_i \in \mathbf{C}$. That is, F is an element of the ring generated by convergent power series, functions $(z-a)^\alpha$ and $\log(z-a)$.

Let $f : U_a' \longrightarrow U_a$ be a covering of U_a. Since the universal covering surface $\tilde{U}_a$ is also a covering of U_a', with projection μ, we can consider $O(U_a') = \mu^*[O(U_a')] \subset O(\tilde{U}_a)$.

Therefore, a function on U_a' can be considered as a function on $\tilde{U}_a$. Thus it makes sense to say that a function on U_a' has a regular singular point at a.

Note here that if the α_i are integers and $B_i = C_i = \cdots = 0$, then F is a convergent power series which is holomorphic and single valued at a. Therefore, the notion of being holomorphic at a is a special case of having a regular singular point at a.

Preparation Theorem 18.1. If $F, G \in O(\tilde{U}_a)$ have regular singular points at a, then

$$F+G, \ F-G, \ \lambda F + \mu G, \ FG, \ \text{and} \ \frac{dF}{dz}$$

also have regular singular points at a.

Preparation Theorem 18.2. If $F, G \in O(\tilde{U}_a)$ have regular singular points at a, and $F/G \in K(U_a)$, then F/G (as a function in U_a) has a pole at a. (That is, a is not an essential singularity.)

In order to prove Preparation Theorem 18.2, we need the following two lemmas.

Lemma 18.3. Let $\alpha_1, \alpha_2, \ldots, \alpha_n$ be complex numbers such that $\alpha_i - \alpha_j$ is not an integer for $i \neq j$. Then, for every positive integer n, the following $(N+1)n$ sequences are linearly independent over $\mathbf{C}$:

$$\{m^k e^{2\pi\sqrt{-1}\alpha_i m}\}_{m=1}^{\infty} \qquad \begin{pmatrix} i = 1, \ldots, n \\ k = 0, \ldots, N \end{pmatrix}$$

In other words, if the equation $\sum_{i=1}^{n} \sum_{k=0}^{N} C_{i,k} m^k e^{2\pi\sqrt{-1}\alpha_i m} = 0$ $(m = 1, 2, \ldots)$ holds for complex numbers $C_{i,k}$, then $C_{i,k} = 0$ $(i = 1, \ldots, n; \; k = 0, \ldots, N)$.

The proof is easy.

For simplicity, let us assume that the center a of U_a is 0. The field of meromorphic functions $K(\tilde{U}_a)$ on the universal covering surface $\tilde{U}_a$ contains the field $K(U_a)$ and the functions $z^{\alpha}(\log z)^k$. We have the following.

Lemma 18.4. Let $\alpha_1, \alpha_2, \ldots, \alpha_n$ be complex numbers such that $\alpha_i - \alpha_j$ is not an integer for $i \neq j$. Then, for every positive integer n, the $(N+1)n$ functions

$$z^{\alpha_i}(\log z)^k \; (i = 1, \ldots, n; \; k = 0, \ldots, N)$$

are linearly independent over $K(U_a)$; namely, if the equation

$$\sum_{i=1}^{n} \sum_{k=0}^{N} F_{i,k}(z) z^{\alpha_i}(\log z)^k \equiv 0$$

holds for functions $F_{i,k}$, then $F_{i,k} \equiv 0$.

Proof: Apply the covering transformation $(\gamma^m)^*$ to both sides of the equation above. Then we have

$$\sum_{i=1}^{n} \sum_{k=0}^{N} F_{i,k}(z) z^{\alpha_i} e^{2\pi\sqrt{-1}\alpha_i m}(\log z + 2\pi\sqrt{-1}m)^k \equiv 0$$

Comparing the coefficients of $m^h e^{2\pi\sqrt{-1}\alpha_i m}$, we obtain

$$\sum_{k=h}^{N} F_{i,k}(z)(\log z)^{k-h}\binom{k}{h} \equiv 0,$$

By Lemma 18.3. Apply this argument again. Then we have

$$\sum_{k=h}^{N} F_{i,k}(z)(\log z + 2\pi\sqrt{-1}m)^{k-h}\binom{k}{h} \equiv 0.$$

Again comparing constant terms, we obtain $F_{i,k}(z) \equiv 0$.   Q.E.D.

Preparation Theorem 18.2 is proved as follows. Let

$$F = \sum_{i=1}^{n}\sum_{k=0}^{N} z^{\alpha_i}(\log z)^k P_{i,k}(z),$$

$$G = \sum_{i=1}^{n}\sum_{k=0}^{N} z^{\alpha_i}(\log z)^k Q_{i,k}(z),$$

where $P_{i,k}, Q_{i,k}$ are convergent power series in z. If we allow zero coefficients in $P_{i,k}, Q_{i,k}$, then we can choose the same $\alpha_1, \ldots, \alpha_n$ in both F and G. We can also assume that $\alpha_i - \alpha_j \notin \mathbf{Z}$ if $i \neq j$, because if $\alpha_j = \alpha_i + m$ $(m \in \mathbf{Z})$, then $z^{\alpha_j} = z^{\alpha_i} z^m$. Then we can rewrite $P_{j,k}(z)$ to include z^m. Now let $f = F/G$. Then $F = fG$, and we have

$$\sum_{i}\sum_{k} z^{\alpha}(\log z)^k (P_{i,k} - fQ_{i,k}) \equiv 0.$$

Since $f \in K(U_a)$, we can assume that $P_{i,k} - fQ_{i,k} \in K(U_a)$ by choosing a smaller U_a, if necessary.

Therefore we have $P_{i,k} - fQ_{i,k} \equiv 0$, by Lemma 18.4. Since $G \neq 0$, there exists a pair (i, k) such that $Q_{i,k} \neq 0$. So $f = \dfrac{P_{i,k}}{Q_{i,k}}$ for this pair. This implies that f is meromorphic at $z = 0$.   Q.E.D.

Consider a differential equation

(#) $$\frac{d^2 w}{dz^2} + P(z)\frac{dw}{dz} + Q(z)w = 0$$

where $P(z)$ and $Q(z)$ are holomorphic functions on U_a. The two-dimensional vector space $V_{\#}$ of solutions of (#) is contained in $O(\tilde{U}_a)$. We say that the differential equation (#) is of *Fuchsian type* at the point a if every solution of (#) has a regular singular point at a. (In particular, a solution may be holomorphic at a.) The following theorem is well known.

Theorem 18.5. The differential equation

(#)
$$\frac{d^2w}{dz^2} + P(z)\frac{dw}{dz} + Q(z)w = 0,$$

where $P, Q \in O(U_a)$, is of Fuchsian type at the point a if and only if $P(z)$ and $Q(z)$ have Laurent expansions at a of the form

$$P(z) = \frac{\alpha_0}{z-a} + \alpha_1 + \alpha_2(z-a) + \alpha_3(z-a)^2 + \cdots$$

$$Q(z) = \frac{\beta_0}{(z-a)^2} + \frac{\beta_1}{(z-a)} + \beta_2 + \beta_3(z-a) + \cdots$$

Here we sketch the "if" direction of the proof. For details, see [4], for example.

Lemma 18.6. Let φ and ψ be linearly independent solutions of (#). Then we have

$$\varphi\psi' - \varphi'\psi = \begin{vmatrix} \varphi & \psi \\ \varphi' & \psi' \end{vmatrix} = Ce^{-\int P(z)dz},$$

$$P(z) = -(\varphi\psi' - \varphi'\psi)'/(\varphi\psi' - \varphi'\psi).$$

Let

$$W = \begin{vmatrix} \varphi & \psi \\ \varphi' & \psi' \end{vmatrix}$$

(the *Wronskian*). Then

$$W' = \begin{vmatrix} \varphi' & \psi' \\ \varphi' & \psi' \end{vmatrix} + \begin{vmatrix} \varphi & \psi \\ \varphi'' & \psi'' \end{vmatrix} = \begin{vmatrix} \varphi & \psi \\ \varphi'' & \psi'' \end{vmatrix} = \begin{vmatrix} \varphi & \psi \\ -P\varphi' - Q\varphi & -P\psi' - Q\psi \end{vmatrix}$$

$$= \begin{vmatrix} \varphi & \psi \\ -P\varphi' & -P\psi' \end{vmatrix} = -P\begin{vmatrix} \varphi & \psi \\ \varphi' & \psi' \end{vmatrix} = -PW.$$

Therefore, W is a solution of $W' = -PW$. Hence

$$\frac{W'}{W} = -P;$$

$$\therefore \qquad \log W = -\int P dz + C;$$

$$\therefore \qquad W = Ce^{-\int P dz}.$$

Thus we have

$$P(z) = -\frac{W'}{W} = -\frac{(\varphi\psi' - \varphi'\psi)'}{(\varphi\psi' - \varphi'\psi)}.$$

(End of proof of Lemma 18.6.)

Since the covering transformation group $\Gamma = \Gamma(\tilde{U}_a \xrightarrow{z} U_a) \cong \pi_1(U_a; b)$ is a cyclic group generated by γ, the monodromy representation of Γ defined by $(\#)$ is determined by the action $w \mapsto \gamma^*(w)$ of the generator γ on w..

Consider the Jordan canonical form of the linear transformation $\gamma \mapsto \gamma^*(w)$ of $V_\#$. It is of the form

$$\begin{pmatrix} c & 0 \\ 0 & d \end{pmatrix} \quad \text{or} \quad \begin{pmatrix} c & 1 \\ 0 & c \end{pmatrix}.$$

Case I. γ^* has Jordan form $\begin{pmatrix} c & 0 \\ 0 & d \end{pmatrix}$.

Then $V_\#$ has a basis $[w_1, w_2]$ such that

$$(\gamma^* w_1, \gamma^* w_2) = (w_1, w_2) \begin{pmatrix} c & 0 \\ 0 & d \end{pmatrix}.$$

This means that

$$\gamma^* w_1 = c w_1, \qquad \gamma^* w_2 = d w_2.$$

Choose $\lambda \in \mathbf{C}$ such that $e^{2\pi i \lambda} = c$, and consider $w_1/(z-a)^\lambda$. Then we have

$$\gamma^*(w_1/(z-a)^\lambda) = (\gamma^* w_1)/\gamma^*(z-a)^\lambda = c w_1/e^{2\pi i \lambda}(z-a)^\lambda$$
$$= c w_1/c(z-a)^\lambda = w_1/(z-a)^\lambda.$$

Therefore, $\dfrac{w_1}{(z-a)^\lambda}$ is a single-valued function on U_a. Let

$$\frac{w_1}{(z-a)^\lambda} = \sum_{v=-\infty}^{\infty} c_v (z-a)^v$$

be its Laurent expansion. Since $(\#)$ is of Fuchsian type at a, w_1 has a regular singular point at a. Hence, by Preparation Theorem 18.2, this function is meromorphic at a. This means that the Laurent series has finite principal part:

$$\frac{w_1}{(z-a)^\lambda} = \sum_{n \geq -v_0} c_n (z-a)^n.$$

Let $\lambda_1 = \lambda - v_0$ and rename c_{n-v_0} by c_n. Then we have

$$w_1 = (z-a)^{\lambda_1} \sum_{n=0}^{\infty} c_n (z-a)^n \qquad (c_0 \neq 0)$$

We also have

$$e^{2\pi i\lambda_1} = e^{2\pi i\lambda} = c.$$

Similarly,

$$w_2 = (z-a)^{\lambda_2} \sum_{n=0}^{\infty} c'_n(z-a)^n \qquad (c'_0 \neq 0)$$

$$e^{2\pi i\lambda_2} = d.$$

Let us compute the Wronskian $W = \begin{vmatrix} w_1 & w_2 \\ w'_1 & w'_2 \end{vmatrix}$. Set $t = z - a$ for simplicity. Since $w_i = t^{\lambda_i} R_i(t)$, where $R_i(t)$ is a convergent power series (c.p.s.) and $R_i(0) \neq 0$, we have

$$w'_i = \lambda_i t^{\lambda_i - 1} \cdot R_i + t^{\lambda_i} R'_i = t^{\lambda_i - 1}(\lambda_i R_i + tR'_i).$$

Then

$$
\begin{aligned}
W &= \begin{vmatrix} t^{\lambda_1} R_1 & t^{\lambda_2} R_2 \\ t^{\lambda_1 - 1}(\lambda_1 R_1 + tR'_1) & t^{\lambda_2 - 1}(\lambda_2 R_2 + tR'_2) \end{vmatrix} \\
&= t^{\lambda_1 + \lambda_2 - 1} \begin{vmatrix} R_1 & R_2 \\ \lambda_1 R_1 + tR'_1 & \lambda_2 R_2 + tR'_2 \end{vmatrix} \\
&= t^{\lambda_1 + \lambda_2 - 1} \{(\lambda_2 - \lambda_1)R_1 R_2 + t(R_1 R'_2 - R'_1 R_2)\}.
\end{aligned}
$$

Therefore, $W = t^{\mu} \cdot R(t)$ where $R(t)$ is a convergent power series, $R(0) \neq 0$. So

$$P = -W'/W = -(\mu t^{\mu-1}R + t^{\mu}R')/(t^{\mu} \cdot R) = -\mu/t - R'/R = -\frac{\mu}{t} + \text{c.p.s.}$$

Thus we know that

$$P(z) = -\frac{\mu}{z-a} + \text{c.p.s.}$$

Because $0 = w''_1 + Pw'_1 + Qw_1$, we have

$$
\begin{aligned}
Q &= -(w''_1 + Pw'_1)/w_1 \\
&= \frac{t^{\lambda_1 - 2} \cdot \text{c.p.s.} + (\frac{\mu}{t} + \text{c.p.s.})t^{\lambda-1} \cdot \text{c.p.s.}}{t^{\lambda_1} \cdot R_1(t)} \\
&= (t^{\lambda_1 - 2} \cdot \text{c.p.s.})/(t^{\lambda_1} \cdot R_1(t)) \\
&= t^{-2} \cdot \text{c.p.s.} /R_1(t), \qquad R_1(0) \neq 0.
\end{aligned}
$$

Thus

$$Q = t^{-2} \cdot (\text{c.p.s.}).$$

So

$$Q(z) = \frac{\beta}{(z-a)^2} + \frac{\delta}{z-a} + \text{c.p.s.}$$

This completes the proof of the "if" part of Theroem 18.5, Case I.

Case II. γ^* has Jordan form $\begin{pmatrix} c & 1 \\ 0 & c \end{pmatrix}$.

We can choose a basis for $V_\#$ so that

$$(\gamma^* w_1, \gamma^* w_2) = (w_1, w_2) \begin{pmatrix} c & 1 \\ 0 & c \end{pmatrix};$$

that is, $\gamma^* w_1 = c w_1$, $\gamma^* w_2 = w_1 + c w_2$. As in Case I, w_1 has the form

$$w_1 = (z - a)^{\lambda_1} \cdot \sum_{n=0}^{\infty} c_n (z - a)^n \qquad (c_0 \neq 0)$$

$$e^{2\pi i \lambda_1} = c.$$

In order to study w_2, let $w_3 = \dfrac{1}{2\pi i c}(\log(z - a)) w_1$. Then

$$\gamma^* w_3 = \frac{1}{2\pi i c} \cdot (\log(z - a) + 2\pi i) c w_1 = w_1 + c w_3.$$

So let $w_4 = w_2 - w_3$. Then

$$\gamma^* w_4 = \gamma^* w_2 - \gamma^* w_3 = w_1 + c w_2 - w_1 - c w_3 = c w_4.$$

Therefore, we have $w_4 = (z - a)^{\lambda_2} R_4(z - a)$. Thus $w_2 = w_4 + w_3$ has the following expansion:

$$w_2 = (z - a)^{\lambda_2} \sum b_n (z - a)^n + (z - a)^{\lambda_1} \log(z - a) \sum d_n (z - a)^n$$

$$e^{2\pi i \lambda_2} = e^{2\pi i \lambda_1} = c$$

$\cdot$

By computing $P = -W'/W$, where $W = \begin{vmatrix} w_1 & w_2 \\ w_1' & w_2' \end{vmatrix}$, and $Q = \dfrac{-w_1'' - P w_1'}{w_1}$, we obtain Laurent expansions as in Case I. This ends the proof of (half of) Theorem 18.5.

As a by-product of the argument, we obtain the following:

Theorem 18.6. Let
$$\frac{d^2w}{dz^2} + P(z)\frac{dw}{dz} + Q(z)w = 0$$
be of Fuchsian type at a, with $P, Q \in O(U_a)$.

(I) If the generator of the monodromy group has Jordan form
$$\begin{pmatrix} c & 0 \\ 0 & d \end{pmatrix},$$
then the solution space $V_\#$ of $(\#)$ is spanned by
$$w_1 = (z-a)^\lambda \sum_{n=0}^\infty c_n(z-a)^n \qquad (c_0 \neq 0)$$
$$w_2 = (z-a)^\mu \sum_{n=0}^\infty d_n(z-a)^n \qquad (d_0 \neq 0)$$
and we have
$$(\gamma^*w_1, \gamma^*w_2) = (w_1, w_2)\begin{pmatrix} c & 0 \\ 0 & d \end{pmatrix},$$
and where λ and μ satisfy $e^{2\pi i\lambda} = c$, $e^{2\pi i\mu} = d$.

(II) If the Jordan form is
$$\begin{pmatrix} c & 1 \\ 0 & c \end{pmatrix},$$
then $V_\#$ is spanned by
$$w_1 = (z-a)^\lambda \sum_{n=0}^\infty c_n(z-a)^n \qquad (c_0 \neq 0)$$
$$w_2 = (z-a)^\mu \sum_{n=0}^\infty d_n(z-a)^n + \frac{1}{2\pi i c}\log(z-a)\cdot w_1, \qquad (d_0 \neq 0)$$
We have
$$(\gamma^*w_1, \gamma^*w_2) = (w_1, w_2)\begin{pmatrix} c & 1 \\ 0 & c \end{pmatrix},$$
and λ and μ satisfy $e^{2\pi i\lambda} = e^{2\pi i\mu} = c$. (Therefore $\lambda - \mu$ is an integer. We will see later that $\lambda - \mu \geq 0$.)

Next, we sketch the proof of the "only if" part of Theorem 18.5. Let $P(z)$

and $Q(z)$ be the Laurent series of Theorem 18.5, and set $A(z) = (z - a)P(z)$, $B(z) = (z - a)^2 Q(z)$. Hence A and B are holomorphic at $z = a$. Let $A(z) = \sum_{n=0}^{\infty} \alpha_n (z - a)^n$, $B(z) = \sum_{n=0}^{\infty} \beta_n (z - a)^n$ be their Taylor expansions. Then (#) becomes

$$(z - a)^2 \frac{d^2 w}{dz^2} + (z - a)A(z)\frac{dw}{dz} + B(z)w = 0.$$

Suppose that (#) has a solution of the form

$$w(z) = (z - a)^{\lambda} \sum_{n=0}^{\infty} c_n (z - a)^n, \qquad c_0 \neq 0.$$

Substitute this series into equation (#). We denote $z - a$ by t. Then

$$w = t^{\lambda} \sum_{n=0}^{\infty} c_n t^n = \sum_{n=0}^{\infty} c_n t^{\lambda+n},$$

$$\frac{dw}{dz} = \sum_{n=0}^{\infty} (\lambda + n)c_n t^{\lambda+n-1},$$

$$\frac{d^2 w}{dz^2} = \sum_{n=0}^{\infty} (\lambda + n)(\lambda + n - 1)c_n t^{\lambda+n-2}.$$

Therefore,

$$0 = t^2 \frac{d^2 w}{dz^2} + tA(z)\frac{dw}{dz} + B(z)w$$

$$= \sum_{n=0}^{\infty} \Big[(\lambda + n)(\lambda + n - 1)c_n + \sum_{k=0}^{n} (\lambda + k)c_k \alpha_{n-k} + \sum_{k=0}^{n} c_k \beta_{n-k} \Big] t^{\lambda+n}$$

So

(1) $[(\lambda + n)(\lambda + n - 1) + \alpha_0(\lambda + n) + \beta_0]c_n + \sum_{k=0}^{n-1} [\alpha_{n-k}(\lambda + k) + \beta_{n-k}]c_k = 0$

$$(n = 0, 1, 2, \dots)$$

In particular, for $n = 0$, since $c_0 \neq 0$, we have

(2) $$\lambda(\lambda - 1) + \alpha_0 \lambda + \beta_0 = 0$$

Then λ is a root of the quadratic equation

(3) $$X(X - 1) + \alpha_0 X + \beta_0 = 0.$$

We call this the equation the *indicial equation* of the differential equation (#) at $z = a$. From (1) we have

$$F(\lambda + n)c_n = -\sum_{k=0}^{n-1} [\alpha_{n-k}(\lambda + k) + \beta_{n-k}]c_k.$$

If the two roots of $F(X) = 0$ do not differ by an integer, then $F(\lambda + n) \neq 0$. Therefore we can define $c_1, c_2, c_3, \ldots$ recursively from c_0 by

$$(4) \qquad c_n = \frac{-1}{F(\lambda + n)} \sum_{k=0}^{n-1} [\alpha_{n-k}(\lambda + k) + \beta_{n-k}] c_k.$$

Conversely, let λ be a root of $F(X) = 0$, and let $c_0 = 1$. Define a power series $\sum c_n t^n$, where the coefficients $c_1, c_2, c_3, \ldots$ are given by the recursion formula (4). Then a simple estimate shows that this sequence converges at $t = 0$, and $t^\lambda \sum c_n t^n$ gives a solution of

$$t^2 \frac{d^2 w}{dt^2} + t A(t + a) \frac{dw}{dt} + B(t + a) w = 0.$$

Since the quadratic equation $F(X) = 0$ has two roots, λ and μ, we have two solutions, $w_1(z) = (z - a)^\lambda \sum c_n (z - a)^n$ and $w_2(z) = (z - a)^\mu \sum c'_n (z - a)^n$, which are obviously linearly independent. Therefore, an arbitrary solution can be written as

$$w(z) = (z - a)^\lambda c \sum c_n (z - a)^n + (z - a)^\mu c' \sum c'_n (z - a)^n.$$

Therefore (#) is of Fuchsian type at $z = a$.

Next let us consider the case in which the two roots of $F(X) = 0$ differ by an integer: $\lambda - \mu = m$. We can assume that $m \geq 0$. As before, we can define a solution

$$w_1(z) = (z - a)^\lambda \sum c_n (z - a)^n$$

from the root λ. Here we may assume that $c_0 = 1$. But we cannot obtain another solution $(z - a)^\mu \sum c'_n (z - a)^n$ from the other root because the denominator $F(\mu + n)$ in the recursive formula

$$(4') \qquad c'_n = \frac{-1}{F(\mu + n)} \sum_{k=0}^{n-1} [\alpha_{n-k}(\mu + k) + \beta_{n-k}] c'_k$$

has zeros when $n = m$. In order to obtain another solution w of (#), we let $w = w_1 \eta$, as usual. Since

$$w' = w'_1 \eta + w_1 \eta'$$
$$w'' = w''_1 \eta + 2 w'_1 \eta' + w_1 \eta'',$$

we have

$$0 = t^2 w'' + t A w' + B w$$
$$= (t^2 w''_1 + t A w'_1 + B w_1) \eta + t^2 (2 w'_1 \eta' + w_1 \eta'') + t A w_1 \eta'.$$

Therefore

$$\therefore \qquad t(2w_1'\eta' + w_1\eta'') + Aw_1\eta' = 0$$

$$\therefore \qquad \eta'' + \left(2\frac{w_1'}{w_1} + \frac{A}{t}\right)\eta' = 0$$

$$\therefore \qquad \log\eta' = -2\log w_1 - \int\frac{A}{t}dt$$

$$\therefore \qquad \eta' = w_1^{-2}e^{-\int\frac{A}{t}dt}$$

$$\therefore \qquad \eta = \int w_1^{-2}e^{-\int\frac{A}{t}dt}dt$$

$$\therefore \qquad w = w_1\eta = w_1\int w_1^{-2}e^{-\int\frac{A}{t}dt}dt$$

The function w is linearly independent from w_1. So,

$$w_1 = t^\lambda \sum c_n t^n \qquad c_0 = 1.$$

$$A = \alpha_0 + \alpha_1 t + \cdots \qquad \alpha_0 \neq 0.$$

$$\therefore \qquad \int\frac{A}{t}dt = \int\left(\frac{\alpha_0}{t} + \alpha_1 + \alpha_2 t + \cdots\right)dt = \alpha_0\log t + \alpha_1 t + \frac{\alpha_2}{2}t^2\cdots$$

$$\therefore \qquad e^{-\int\frac{A}{t}dt} = t^{-\alpha_0}e^{-(\alpha_1 t + \frac{\alpha_2}{2}t^2 + \cdots)} = t^{-\alpha_0}(1 + b_1 t + b_2 t^2 + \cdots).$$

On the other hand,

$$\therefore \qquad w_1^{-2} = \{t^\lambda(1 + c_1 t + \cdots)\}^{-2} = t^{-2\lambda}(1 + d_1 t + d_2 t^2 + \cdots).$$

$$\therefore \qquad w_1^{-2}e^{-\int\frac{A}{t}dt} = t^{-\alpha_0-2\lambda}(1 + e_1 t + e_2 t^2 + \cdots).$$

Since λ and $\mu = \lambda - m$ are two roots of $F(X) = X(X-1) + \alpha_0 X + \beta_0 = 0$, we have the relation $\lambda + \mu = 2\lambda - m = 1 - \alpha_0$; so $-\alpha_0 - 2\lambda = -1 - m$. Therefore,

$$w_1^{-2}e^{-\int\frac{A}{t}dt} = t^{-m-1}(1 + e_1 t + \cdots).$$

Hence,

$$\eta = \int w_1^{-2}e^{-\int\frac{A}{t}dt}dt$$

$$= \frac{t^{-m}}{-m} + \frac{e_1 t^{-m+1}}{-m+1} + \cdots + e_m\log t + e_{m+1}t + \frac{e_{m+2}t^2}{2} + \cdots + C,$$

where C is a constant of integration. Thus we have

$$w = w_1\eta$$

$$= e_m\log t \cdot w_1 + \left(\sum_{n=0}^\infty t^\lambda c_n t^n\right)\left(\frac{t^{-m}}{-m} + \cdots\right)$$

$$= e_m(\log t)w_1 + t^\mu\sum_{n=0}^\infty c_n' t^n$$

$$= e_m(z-a)^\lambda\log(z-a)\sum_{n=0}^\infty c_n(z-a)^n + (z-a)^\mu\sum_{n=0}^\infty c_n'(z-a)^n$$

The function w has a regular singular point at $z = a$. Thus (#) is of Fuchsian type.

Remark. We may have $e_m = 0$. Then $\log(z - a)$ does not appear above.

As a byproduct of the above argument, we have

Theorem 18.8.   The exponents λ and μ in Theorem 18.7 are roots of the equation $F(X) = 0$, where $F(X) = X(X - 1) + \alpha_0 X + \beta_0$.

The Nineteenth Week:
Fuchsian Differential Equations

[A] Let D be the region obtained by removing the points $a_1, a_2, ..., a_n$ from the complex plane: $D = \mathbf{C} - \{a_1, a_2, ..., a_n\}$. It can also be obtained by removing the $n+1$ points $\{a_1, a_2, ..., a_n, a_{n+1} = \infty\}$ from the Riemann sphere $\mathbf{C} \cup \{\infty\}$. Let $z : \tilde{D} \to D$ be the universal covering surface of D. We take a "5-yen coin" U_{a_i} around each a_i (defined at the beginning of Week 18). Let $\tilde{U}_{a_i,1}, \tilde{U}_{a_i,2}, \tilde{U}_{a_i,3}, ...$ be the connected components of the open subset $z^{-1}(U_{a_i})$ of $\tilde{D}$. Then each $z : \tilde{U}_{a_i,j} \longrightarrow U_{a_i}$ is a covering of U_{a_i}. We call it a *spiral staircase* which covers U_{a_i}.

For a function $F \in O(\tilde{D})$, we say that F has *regular singular points* at $a_1, a_2, ..., a_n$ if the restriction $F|_{\tilde{U}_{a_i,j}}$ of F to each spiral staircase $\tilde{U}_{a_i,j}$ ($i = 1, 2, ...; j = 1, 2, ...$) has a regular singular point at a_i. We say that $F(z)$ has a regular singular point at $z = \infty$ if $H(t) = F(\frac{1}{t})$ has a regular singular point at $t = 0$.

Theorem 19.1. If $F, G \in O(\tilde{D})$, ($G \not\equiv 0$) have regular singular points at $a_1, a_2, ..., a_n, a_{n+1} = \infty$, and F/G is in $K(D)$ ($= z^*(K(D)) \subset K(\tilde{D})$), then F/G is a rational function of z.

Proof: Of course, F/G is meromorphic on D. We also know that it is meromorphic at $a_1, a_2, ..., a_n, \infty$ by Preparation Theorem 18.2. Therefore it is meromorphic on the entire Riemann sphere. Hence it is rational, by a well-known theorem in complex analysis. Q.E.D.

Let $B(\tilde{D})$ be the set of meromorphic functions on $\tilde{D}$ having regular singular points at $a_1, a_2, ..., a_n, \infty$. Then $B(\tilde{D})$ is a ring, and $B(\tilde{D}) \subset O(\tilde{D})$; let $M(\tilde{D})$ denote the field of quotients of $B(\tilde{D})$. We denote by $K = K(R) = \mathbf{C}(z)$ the set of rational functions on the Riemann sphere R. Then the above theorem can be summarized as

$$M(\tilde{D}) \cap K(D) \subset K(R) = \mathbf{C}(z).$$

[B] Consider the differential equation

(#)
$$\frac{d^2w}{dz^2} + P(z)\frac{dw}{dz} + Q(z)w = 0.$$

(a) We assume that $P(z)$ and $Q(z)$ are rational, and that their poles are contained in the set $\{a_1, a_2, ..., a_n\}$. Let $D = \mathbf{C} - \{a_1, a_2, ..., a_n\}$. A solution of (#) is a function defined on $\tilde{D}$; that is, $V_{\#} \subset O(\tilde{D})$. Furthermore, we impose the following condition:

(b) Every solution of (#) has regular singular points in the set $\{a_1, a_2, ..., a_n, \infty\}$.

The differential equation (#) is said to be of *Fuchsian type* if (a) and (b) are satisfied. In short, a Fuchsian differential equation is a differential equation whose coefficients are rational functions, and whose solutions have only regular singular points.

The following is a famous theorem.

Theorem 19.2. The differential equation (#) is of Fuchsian type if and only if $P(z)$ and $Q(z)$ are rational functions of the following form:

(*)
$$P(z) = \sum_{i=1}^{n} \frac{\alpha_i}{z - a_i} = \frac{\text{polynomial in } z \text{ of degree } \leq n - 1}{(z - a_1) \cdots (z - a_n)}.$$

(**)
$$Q(z) = \sum_{i=1}^{n} \left\{ \frac{\beta_i}{(z - a_i)^2} + \frac{\delta_i}{z - a_i} \right\}, \quad \text{where} \quad \sum_{i=1}^{n} \delta_i = 0$$
$$= \frac{\text{polynomial in } z \text{ of degree } \leq 2(n - 1)}{[(z - a_1) \cdots (z - a_n)]^2}.$$

Proof: (We give only the "if" part.) Assume that P and Q are rational functions satisying (*) and (**). Let us expand $P(z)$ and $Q(z)$ into partial fractions. Since $a_1, a_2, ..., a_n$ are the only poles of $P(z)$ and $Q(z)$, and (#) is Fuchsian at each a_i, we have

(1)
$$P(z) = \sum_{i=1}^{n} \frac{\alpha_i}{z - a_i} + D(z)$$
$$Q(z) = \sum_{i=1}^{n} \left(\frac{\beta_i}{(z - a_i)^2} + \frac{\delta_i}{z - a_i} \right) + E(z),$$

where D and E are polynomials in z, by Theorem 18.5. Let us find a condition for (#) to be Fuchsian at $z = \infty$, which is what we need to prove. Let $z = \frac{1}{t}$. Then

$$\frac{dw}{dz} = \frac{dw}{dt} \frac{-1}{z^2} = -t^2 \frac{dw}{dt}$$
$$\frac{d^2w}{dz^2} = \frac{2}{z^3} \frac{dw}{dt} + \frac{-1}{z^2} \frac{d}{dz} \left(\frac{dw}{dt} \right) = 2t^3 \frac{dw}{dt} + t^4 \frac{d^2w}{dt^2}.$$

Therefore, (#) becomes

$$0 = \frac{d^2w}{dz^2} + P\frac{dw}{dz} + Qw = t^4\frac{d^2w}{dt^2} + 2t^3\frac{dw}{dt} - Pt^2\frac{dw}{dt} + Qw,$$

and hence

(♮)
$$\frac{d^2w}{dt^2} + \left(\frac{2}{t} - \frac{1}{t^2}P\left(\frac{1}{t}\right)\right)\frac{dw}{dt} + \frac{1}{t^4}Q\left(\frac{1}{t}\right)w = 0$$

For (♮) to be of Fuchsian type at $t = 0$, we must have

(2)
$$\frac{1}{t^2}P\left(\frac{1}{t}\right) = \frac{1}{t} \cdot \text{c.p.s. in } t$$

$$\frac{1}{t^4}Q\left(\frac{1}{t}\right) = \frac{1}{t^2} \cdot \text{c.p.s. in } t$$

Using (1), we have

$$\frac{1}{t^2}P\left(\frac{1}{t}\right) = \left(\sum_{i=1}^{n}\frac{\alpha_i}{1 - a_it}\right)\frac{1}{t} + \frac{1}{t^2}D\left(\frac{1}{t}\right),$$

$$\frac{1}{t^4}Q\left(\frac{1}{t}\right) = \left(\sum_{i=1}^{n}\frac{\beta_i}{(1 - a_it)^2}\right)\frac{1}{t^2} + \left(\sum_{i=1}^{n}\frac{\delta_i}{1 - a_it}\right)\frac{1}{t^3} + \frac{1}{t^4}E\left(\frac{1}{t}\right).$$

In order to show (2), set $D = E \equiv 0$, $\sum_{i=1}^{n}\delta_i = 0$. We can expand $\frac{1}{1 - a_it} = \sum_{j=0}^{\infty}(a_it)^j$. Then

$$\frac{1}{t}P\left(\frac{1}{t}\right) = \frac{1}{t} \cdot \text{c.p.s. in } t,$$

as desired. The coefficient of $\frac{dw}{dt}$ in (♮) is

(3)
$$\frac{2}{t} - \frac{1}{t^2}P\left(\frac{1}{t}\right) = \frac{1}{t}\left(2 - \sum_{i=1}^{n}\alpha_i\right) + \text{c.p.s. in } t.$$

Using the same expansions in $\frac{1}{t^4}Q\left(\frac{1}{t}\right)$, the second term becomes

$$\frac{1}{t^3}\sum_{i}\delta_i\sum_{j}(a_it)^j = \frac{1}{t^3}\left(\sum_{i}\delta_i + \sum_{i}\delta_i\sum_{j=1}^{\infty}(a_it)^j\right)$$

$$= \frac{1}{t^2}\sum_{i}\delta_i a_i\sum_{j=0}^{\infty}(a_it)^j$$

Thus the coefficient of w in (♮) is

(4)
$$\frac{1}{t^4} Q\left(\frac{1}{t}\right) = \frac{1}{t^2} \sum_i \beta_i \left(\sum_{j=0}^{\infty} (a_i t)^j\right)^2 + \frac{1}{t^2} \sum_i \delta_i a_i \sum_{j=0}^{\infty} (a_i t)^j$$

$$= \frac{1}{t^2} \sum_i (\beta_i + \delta_i a_i) + \frac{1}{t} \cdot \text{c.p.s. in } t.$$

Q.E.D.

As we have seen above, every solution $w(z)$ of (#) gives a solution of (♮) by setting $t = \frac{1}{z}$. Using the coefficients of $\frac{1}{t}$ and $\frac{1}{t^2}$ in (3) and (4), we obtain the indicial equation of (♮) at $t = 0$:

(5)
$$X(X-1) + \alpha_\infty X + \beta_\infty = 0$$

$$\alpha_\infty = 2 - \sum_{i=1}^{n} \alpha_i, \quad \beta_\infty = \sum_{i=1}^{n} (\beta_i + a_i \delta_i).$$

We call it the indicial equation of (#) at $z = \infty$. Let λ_∞ and μ_∞ be the two roots of (5).

Problem: Consider a Fuchsian differential equation (#). Let λ_i and μ_i ($i = 1, \ldots, n$) be the roots of the indicial equation $X(X-1) + \alpha_i X + \beta_i = 0$ at each a_i, and let λ_∞ and μ_∞ be the roots of $X(X-1) + \alpha_\infty X + \beta_\infty = 0$, where $\alpha_\infty = \sum_{i=1}^{n} \alpha_i$ and $\beta_\infty = \sum_{i=1}^{n} (\beta_i + a_i \delta_i)$. Prove the Fuchs relation

(6)
$$\sum_{i=1}^{n} (\lambda_i + \mu_i) + \lambda_\infty + \mu_\infty = n - 1.$$

[C] From now on, let $P(z)$ and $Q(z)$ be the functions given in (∗) and (∗∗) and the Fuchsian equation (#). Let $D = \mathbf{C} - \{a_1, a_2, \ldots, a_n\}$. Let $S_\#$ be the field obtained by adjoining all the solutions of (#) and their first derivatives to the field of rational functions $K(R) = \mathbf{C}(z)$:

$$S_\# = \mathbf{C}(z)\left(\left\{w, \frac{dw}{dz}; w \in V_\#\right\}\right).$$

If we choose a pair ϕ, ψ of linearly independent solutions of (#), then the above field is obtained by adjoining $\phi, \psi, \frac{d\phi}{dz},$ and $\frac{d\psi}{dz}$ to $\mathbf{C}(z)$:

$$S_\# = \mathbf{C}(z)\left(\phi, \frac{d\phi}{dz}, \psi, \frac{d\psi}{dz}\right).$$

Since (#) is of Fuchsian type, $\phi, \psi, \frac{d\phi}{dz},$ and $\frac{d\psi}{dz}$ have only regular singular points. Therefore, by Preparation Theorem 18.2, we have $S_\# \cap K(D) \subset \mathbf{C}(z)$. On the other hand, since $S_\# \cap K(D) \supset \mathbf{C}(z)$ is obvious, we have

$$S_\# \cap K(D) = \mathbf{C}(z).$$

Now let us investigate whether (#) is of type L_0 or L over $K = \mathbf{C}(z)$.

Theorem 19.3. For a Fuchsian differential equation (#), every solution of (#) is of type L_0 over $\mathbf{C}(z) = K(R) = K$ if and only if the monodromy representation of (#) is triangulable.

Proof: We omit the "only if" part because it is hard. Let us give the "if" part, which is similar to the argument presented in the 17th Week. Since the monodromy representation of $\Gamma = \pi_1(D)$ is triangulable, there is a basis $[w_1, w_2]$ of $V_\#$ such that

$$\left((\gamma^{-1})^* w_1, (\gamma^{-1})^* w_2 \right) = (w_1, w_2) \begin{pmatrix} a(\gamma) & b(\gamma) \\ 0 & d(\gamma) \end{pmatrix}.$$

In particular, we have

(a) $$(\gamma^{-1})^* w_1 = a(\gamma) w_1.$$

By differentiating both sides, we obtain

(b) $$(\gamma^{-1})^* \left(\frac{dw_1}{dz} \right) = a(\gamma) \frac{dw_1}{dz},$$

and hence

$$(\gamma^{-1})^* \left(\frac{dw_1}{dz} / w_1 \right) = \frac{dw_1}{dz} / w_1 \quad (\forall \gamma \in \Gamma).$$

Therefore $\frac{dw_1}{dz} / w_1 \in K(D)$. On the other hand, since $w_1 \in V_\#$, $\frac{dw_1}{dz} / w_1 \in S_\#$. If we set $A = \frac{dw_1}{dz} / w_1$, then

(c) $$A = \frac{dw_1}{dz} / w_1 \in S_\# \cap K(D) = \mathbf{C}(z).$$

That is, $A = A(z)$ is a rational function of z. From (c), we have $w_1 = Ce^{\int A(z)dz}$, which is of type L_0 over $\mathbf{C}(z)$. Therefore, by Preparation Theorem 17.1, every solution of (#) is of type L_0 over $\mathbf{C}(z)$.   Q.E.D.

[D] Now let us try to explicitly list all of the solutions of (#), assuming that (#) is of type L_0 over $\mathbf{C}(z)$. To this end, it suffices to determine the rational function $A(z)$ of section [C], because the general solution can be written as

$$w = Ce^{\int Adz} \int e^{-2\int Adz - \int Pdz} dz + C'e^{\int Adz},$$

by Preparation Theorem 17.1.

In order to study $A(z)$, let us look at w_1 first. The function w_1 satisfies

$$w_1 \circ \gamma = a(\gamma)w_1 \qquad (\forall \gamma \in \Gamma).$$

Therefore, if $\tilde{p} \in \tilde{D}$ is a zero of w_1, then its conjugate points $\gamma(\tilde{p})$ $(\gamma \in \Gamma)$ are all zeros of w_1. Hence there exist $b_1, \ldots, b_s \in D$ such that the set of zeros of w_1 in $\tilde{D}$ can be written as

$$\{\tilde{p} \in \tilde{D} \mid z(\tilde{p}) = b_1, b_2, \ldots, b_s\}.$$

Since w_1 is a solution of a second order linear differential equation, each of its zeros has multiplicity one. For, if w_1 has $\tilde{p}_1 \in \tilde{D}$ as a zero of multiplicity at least two, then $w \equiv w_1$ and $w \equiv 0$ both satisfy the differential equation (#) with initial conditions $w(\tilde{p}_1) \equiv 0$, $\dfrac{dw}{dz}(\tilde{p}_1) \equiv 0$. Therefore, $w_1 \equiv 0$, by the uniqueness theorem. Since the multiplicity of each zero of w_1 is one, the Laurent expansion of $A(z) = \dfrac{dw_1}{dz}/w_1$ at $z = b_i$ is

(7) $$A(z) = \frac{1}{z - b_i} + \{(z - b_i) \cdot \text{c.p.s. in } (z - b_i)\},$$

with constant term 1. Let λ_i and μ_i be the roots of the indicial equation of (#) at a_i, and let λ_∞ and μ_∞ be the roots of the indicial equation of (#) at ∞. Since w_1 satisfies $w_1 \circ \gamma = a(\gamma)w_1$ $(\forall \gamma \in \Gamma)$, w_1 has an expansion

$$w_1 = (z - a_i)^{\lambda_i} \sum_{n=0}^{\infty} c_n (z - a_i)^n$$

or

$$w_1 = (z - a_i)^{\mu_i} \sum_{n=0}^{\infty} c_n (z - a_i)^n$$

at every a_i. Here, $a(\gamma_i)$ is either $e^{2\pi\sqrt{-1}\lambda_i}$ or $e^{2\pi\sqrt{-1}\mu_i}$, where γ_i winds once around a_i. Similarly, around $z = \infty$, setting $t = \frac{1}{z}$, we have

$$w_1 = t^{\lambda_\infty} \sum c_n t^n \quad \text{or} \quad w_1 = t^{\mu_\infty} \sum c_n t^n,$$

with $a(\gamma_\infty)$ either $e^{2\pi\sqrt{-1}\lambda_\infty}$ or $e^{2\pi\sqrt{-1}\mu_\infty}$. Therefore, at $z = a_i$, we have

$$A(z) = \frac{\dfrac{dw_1}{dz}}{w_1} = \begin{cases} \dfrac{\lambda_i}{z - a_i} + (z - a_i) \cdot \text{c.p.s. in } (z - a_i) \\ \qquad\qquad \text{or} \\ \dfrac{\mu_i}{z - a_i} + (z - a_i) \cdot \text{c.p.s. in } (z - a_i) \end{cases}$$

Similarly, at $z = \infty$, we have

(8) $$A(z) = \frac{\dfrac{dw_1}{dz}}{w_1} = \frac{-t^2 \dfrac{dw_1}{dt}}{w_1} = \begin{cases} -\lambda_\infty t + \cdots \\ -\mu_\infty t + \cdots \end{cases}.$$

Therefore,

$$A(z) = \sum_{i=1}^{n} \frac{\rho_i}{z - a_i} + \sum_{i=1}^{s} \frac{1}{z - b_i}, \quad \text{where } \sum_{i=1}^{n} \rho_i + s = -\rho_\infty.$$

Here, ρ_i represents λ_i or μ_i, and ρ_∞ represents λ_∞ or μ_∞, whichever is appropriate. On the other hand, since $w_1 = e^{\int A(z)dz}$ is a solution of (#), we have

$$\frac{dw_1}{dz} = A(z)w_1,$$

$$\frac{d^2 w_1}{dz^2} = \frac{dA(z)}{dz} w_1 + A(z)\frac{dw_1}{dz} = \frac{dA}{dz} w_1 + A(z)^2 w_1,$$

and hence

$$0 = \frac{d^2 w_1}{dz^2} + P(z)\frac{dw_1}{dz} + Q(z)w_1 = \left\{ \frac{dA}{dz} + A(z)^2 + P(z)A(z) + Q(z) \right\} w_1.$$

Thus

(9) $$\frac{dA}{dz} + A(z)^2 + P(z)A(z) + Q(z) = 0.$$

For simplicity, let $b_j = a_{n+j}$, $\rho_{n+j} = 1$ $(j = 1, \ldots, s)$; $n + s = m$. Then

$$A(z) = \sum_{i=1}^{m} \frac{\rho_i}{z - a_i}.$$

Hence

$$\frac{dA}{dz} = \sum_{i=1}^{m} \frac{-\rho_i}{(z - a_i)^2}.$$

So

$$A(z)^2 = \sum_{i=1}^{m} \frac{\rho_i^2}{(z - a_i)^2} + \sum_{i \neq j} \frac{\rho_i \rho_j}{(z - a_i)(z - a_j)}$$

$$= \sum_{i=1}^{m} \frac{\rho_i^2}{(z - a_i)^2} + \sum_{i \neq j} \left(\frac{1}{z - a_i} - \frac{1}{z - a_j} \right) \frac{\rho_i \rho_j}{a_i - a_j}$$

$$= \sum_{i=1}^{m} \left\{ \frac{\rho_i^2}{(z - a_i)^2} + \frac{2}{z - a_i} \left(\sum_{j \neq i} \frac{\rho_i \rho_j}{a_i - a_j} \right) \right\}$$

If we set $\alpha_{n+j} = 0$, $\beta_{n+j} = 0$, $\delta_{n+j} = 0$ $(j = 1, \ldots, s)$, then we have

$$
\begin{aligned}
P(z)A(z) &= \sum_{i=1}^{m} \frac{\alpha_i}{z - a_i} \sum_{i=1}^{m} \frac{\rho_i}{z - a_i} \\
&= \sum_{i=1}^{m} \frac{\alpha_i \rho_i}{(z - a_i)^2} + \sum_{i \neq j} \frac{\alpha_i \rho_j}{(z - a_i)(z - a_j)} \\
&= \sum_{i=1}^{m} \frac{\alpha_i \rho_i}{(z - a_i)^2} + \sum_{i \neq j} \left(\frac{1}{z - a_i} - \frac{1}{z - a_j} \right) \frac{\alpha_i \rho_j}{a_i - a_j} \\
&= \sum_{i=1}^{m} \left\{ \frac{\alpha_i \rho_i}{(z - a_i)^2} + \frac{1}{z - a_i} \sum_{j \neq i} \frac{\alpha_i \rho_j + \alpha_j \rho_i}{a_i - a_j} \right\}, \\
Q(z) &= \sum_{i=1}^{m} \left\{ \frac{\beta_i}{(z - a_i)^2} + \frac{\delta_i}{z - a_i} \right\}.
\end{aligned}
$$

Thus (9) becomes

(10)
$$
\begin{aligned}
0 = \sum_{i=1}^{m} \Bigg[&\frac{1}{(z - a_i)^2} (-\rho_i + \rho_i^2 + \alpha_i \rho_i + \beta_i) \\
&+ \frac{1}{z - a_i} \left\{ \sum_{j \neq i} \frac{1}{a_i - a_j} (2\rho_i \rho_j + \rho_i \alpha_j + \rho_j \alpha_i) + \delta_i \right\} \Bigg].
\end{aligned}
$$

Therefore,

(11)
$$
\rho_i^2 - \rho_i + \alpha_i \rho_i + \beta_i = 0, \quad (i = 1, \ldots, m)
$$

(12)
$$
\sum_{j \neq i} \frac{1}{a_i - a_j} (2\rho_i \rho_j + \alpha_i \rho_j + \alpha_j \rho_i) + \delta_i = 0, \quad (i = 1, \ldots, m).
$$

Equation (11) says that ρ_i is a root of the indicial equation. We can regard (12) as an algebraic equation for $b_k = a_{n+k}$ $(k = 1, \ldots, s)$ with known coefficients a_i $(i = 1, \ldots, m)$, ρ_j, α_j.

We can determine $A(z) = \sum_{i=1}^{n} \frac{\rho_i}{z - a_1} + \sum_{i=1}^{s} \frac{1}{z - b_i}$ by solving the above equation for $b_1, \ldots, b_s$.

(Summary) How to determine $A(z)$: Let $\lambda_1, \mu_1; \lambda_2, \mu_2; \lambda_n, \mu_n; \ldots; \lambda_\infty, \mu_\infty$ be the roots of the indicial equations of (#) at $a_1, \ldots, a_n, \infty$, respectively. Choose one of λ_i and μ_i and denote it by ρ_i. There are 2^{n+1} ways of choosing $\rho_1, \ldots, \rho_n, \rho_\infty$. We use only those ways satisfying $\sum_{i=1}^{n} \rho_i + \rho_\infty = -s \leq 0$. We can ignore the rest because if $\sum \rho_i + \rho_\infty$ is positive for some choice of $\rho_1, \ldots, \rho_n, \rho_\infty$, then $A(z)$ does not exist, and (#) cannot be of type L_0 over $\mathbf{C}(z)$.

For each $(n+1)$-tuple $(\rho_1, \ldots, \rho_n, \rho_\infty)$ satisfying the condition above, let $s = -(\sum \rho_i + \rho_\infty)$. Recall that for $j = 1, \ldots, s$, $a_{n+j} = b_j$, $\rho_{n+j} = 1$, $\alpha_{n+j} = \beta_{n+j} = \delta_{n+j} = 0$. Then (12) becomes

$$(13) \quad 0 = \delta_i + \sum_{j=i+1}^{n} \frac{1}{a_i - a_j}(2\rho_i\rho_j + \alpha_i\rho_j + \alpha_j\rho_i) + \left(\sum_{k=1}^{s} \frac{1}{a_i - x_k}\right)(2\rho_i + \alpha_i)$$

for $i = 1, 2, \ldots, n$, and

$$(14) \qquad 0 = \sum_{j=1}^{n} \frac{1}{x_i - a_j}(2\rho_j + \alpha_j) + \sum_{\substack{k=1 \\ k \neq i}}^{s} \frac{2}{x_i - x_k}$$

for $i = 1, 2, \ldots, s$. If (13) and (14) have a solution $(x_1, x_2, \ldots, x_s) = (b_1, b_2, \ldots, b_s)$, then

$$A(z) = \sum_{i=1}^{n} \frac{\rho_i}{z - a_i} + \sum_{k=1}^{s} \frac{1}{z - b_k}$$

gives the desired function. But if (13) and (14) have no solution, then $A(z)$ does not exist, and again (#) is not of type L_0 over $\mathbf{C}(z)$. Thus, if we can define $A(z)$, then

$$(15) \qquad \begin{aligned} w_1(z) &= e^{\int A(z)dz} \\ &= e^{\sum \rho_i \log(z - a_i) + \sum \log(z - b_k)} \\ &= \prod_{i=1}^{n}(z - a_i)^{\rho_i} \cdot \prod_{k=1}^{s}(z - b_k) \end{aligned}$$

and the general solution is given by

$$(16) \qquad \begin{aligned} w &= Cw_1 \int w_1^{-2} e^{-\int P dz} dz + C'w_1 \\ &= Cw_1 \int w_1^{-2} e^{-\sum \alpha_i \log(z - a_i)} dz + C'w_1 \\ &= C\prod_{i=1}^{n}(z - a_i)^{\rho_i} \prod_{i=1}^{n}(z - b_k) \cdot \\ &\quad \int \prod_{i=1}^{s}(z - a_i)^{-2\rho_i - \alpha_i} \prod_{k=1}^{s}(z - b_k)^{-2} dz + C'w_1. \end{aligned}$$

[E] Exercise. Solve:

$$\frac{d^2w}{dz^2} + \left\{\frac{1}{3z} + \frac{1}{6(z-1)}\right\}\frac{dw}{dz} + \left\{-\frac{1}{3z^2} - \frac{1}{6(z-1)^2} + \frac{1}{2z(z-1)}\right\}w = 0.$$

Solution: Let us see whether it is of type L_0. Set

$$P(z) = \frac{1}{3z} + \frac{1}{6(z-1)},$$

$$Q(z) = -\frac{1}{3z^2} - \frac{1}{6(z-1)^2} + \frac{1}{2z(z-1)} = -\frac{1}{3z^2} - \frac{1}{6(z-1)^2} - \frac{1}{2z} + \frac{1}{2(z-1)}.$$

The singular points are $0, 1$, and ∞. Because

$$\alpha_0 = \tfrac{1}{3} \qquad\qquad \beta_0 = -\tfrac{1}{3} \qquad\qquad \delta_0 = -\tfrac{1}{2}$$
$$\alpha_1 = \tfrac{1}{6} \qquad\qquad \beta_1 = -\tfrac{1}{6} \qquad\qquad \delta_1 = \tfrac{1}{2}$$
$$\alpha_\infty = 2 - \alpha_1 - \alpha_0 = \tfrac{3}{2} \qquad \beta_\infty = \beta_0 + \beta_1 + \delta_1 = 0$$

the indicial equations are as follows:

Singularity	Indicial equation	Roots λ, μ
0	$X(X-1) + \tfrac{1}{3}X - \tfrac{1}{3} = 0$	$-\tfrac{1}{3}, 1$
1	$X(X-1) + \tfrac{1}{6}X - \tfrac{1}{6} = 0$	$-\tfrac{1}{6}, 1$
∞	$X(X-1) + \tfrac{3}{2}X = 0$	$-\tfrac{1}{2}, 0$

Let $\lambda_0 = -\tfrac{1}{3}, \mu_0 = 1$; $\lambda_1 = -\tfrac{1}{6}, \mu_1 = 1$; $\lambda_\infty = -\tfrac{1}{2}, \mu_\infty = 0$. Set $\delta_i = \lambda_i$ or μ_i. The only combination satisfying $\delta_0 + \delta_1 + \delta_\infty \leq 0$ is $\delta_0 = \lambda_0, \delta_1 = \lambda_1, \delta_\infty = \lambda_\infty$:

$$\lambda_0 + \lambda_1 + \lambda_\infty = -\frac{1}{3} - \frac{1}{6} - \frac{1}{2} = -1,$$

hence $s = 1$. Therefore, if the equation is of type L_0, then it has a solution

$$w_1(z) = e^{\int A(z)dz} = z^{-\frac{1}{3}}(z-1)^{-\frac{1}{6}}(z - C),$$

where

$$A(z) = \frac{-\frac{1}{3}}{z} + \frac{-\frac{1}{6}}{z-1} + \frac{1}{z-C}.$$

Since $A(z)$ satisfies $A' + A^2 + AP + Q = 0$, we can determine C from this condition and we see that $w_1(z)$ is actually a solution. This condition is equivalent to (13) and (14), which say, in our case,

$$0 = -\frac{1}{2} + \frac{1}{0-1}\left\{2(-\frac{1}{3})(-\frac{1}{6}) + \frac{1}{3}(-\frac{1}{6}) + \frac{1}{6}(-\frac{1}{3})\right\}$$
$$+\frac{1}{0-C}\left\{2(-\frac{1}{3}) + \frac{1}{3}\right\},$$

$$0 = \frac{1}{2} + \frac{1}{1-0}\left\{2(-\frac{1}{3})(-\frac{1}{6}) + \frac{1}{6}(-\frac{1}{3}) + \frac{1}{3}(-\frac{1}{6})\right\}$$
$$+\frac{1}{1-C}\left\{2(-\frac{1}{6}) + \frac{1}{6}\right\},$$

$$0 = \frac{1}{C}\left\{2(-\frac{1}{3}) + \frac{1}{3}\right\} + \frac{1}{C-1}\left\{2(-\frac{1}{6}) + \frac{1}{6}\right\}$$

Actually, these are all equivalent, and we get $C = \frac{2}{3}$. Therefore our equation is of type L_0, and has a solution of the form $w_1(z) = z^{-\frac{1}{3}}(z-1)^{-\frac{1}{6}}(z-\frac{2}{3})$. We can check this by direct computation. The other solution is given by

$$w_2(z) = w_1(z) \int w_1(z)^{-2} e^{-\int P(z)dz} dz$$

$$= z^{-\frac{1}{3}}(z-1)^{-\frac{1}{6}}(z-\frac{2}{3}) \int z^{\frac{1}{3}}(z-1)^{\frac{1}{6}}(z-\frac{2}{3})^{-2} dz.$$

References

[1] L. Ahlfors, *Complex Analysis*, 3rd ed., McGraw-Hill, 1979.

[2] F. Ayres, Jr., *Matrices* (Schaum Outline Series), McGraw-Hill, 1962.

[3] F. Ayres, Jr., *Modern Algebra* (Schaum Outline Series), McGraw-Hill, 1965.

[4] G. Birkhoff and G.-C. Rota, *Ordinary Differential Equations*, 4th ed., Wiley, 1989.

[5] C.W. Curtis, *Linear Algebra: An Introductory Approach* (Undergraduate Texts in Mathematics), Springer-Verlag, 1984.

[6] H.M. Farkas and I. Kra, *Riemann Surfaces* (Graduate Texts in Mathematics), Springer-Verlag, 1980.

[7] J. Gray, *Linear Differential Equations and Group Theory from Riemann to Poincaré*, Birkhäuser, 1986.

[8] E.L. Ince, *Ordinary Differential Equations*, Dover.

[9] S. Lipschutz, *General Topology* (Schaum Outline Series), McGraw-Hill, 1965.

[10] S. Lipschutz, *Linear Algebra* (Schaum Outline Series), McGraw-Hill, 1968.

[11] L.S. Pontryagin, *Topological Groups*, 2nd ed., Gordon and Breach, 1966.

[12] I.M. Singer and J.A. Thorpe, *Lecture Notes on Elementary Topology and Geometry* (Undergraduate Texts in Mathematics), Springer-Verlag, 1976.

[13] G Springer, *Introduction to Riemann Surfaces*, 2nd ed., Chelsea, 1981.

[14] M.R. Spiegel, *Complex Variables* (Schaum Outline Series), McGraw-Hill, 1964.

[15] B.L. van der Waerden, *Algebra*, vols. I, II, 4th ed., Ungar, 1970.

[16] M. Yoshida, *Fuchsian Differential Equations*, Friedr. Vieweg & Sohn, 1987.

Notation

$\ni, \in$	element of, 5
$\mathbf{N}$	natural numbers, 5
$\mathbf{Z}$	integers, 5
$\mathbf{Q}$	rational numbers, 6
$\mathbf{R}$	real numbers, 6
$\mathbf{C}$	complex numbers, 6
$\lvert \mathcal{M} \rvert$	cardinality of a set, 8
$\emptyset$	empty set, 8
$\forall$	for all, 8
$\exists$	exists, 8
$\vert$	such that, 8
$\{\dots\}$	set, 9
$\subset, \supset$	subset, 9
$\underset{\neq}{\subset}, \underset{\neq}{\supset}$	proper subset, 10
$\cup, \bigcup_i$	union, 11
$\cap, \bigcap_i$	intersection, 12
$\mapsto$	maps to, 13
$f(\mathcal{M})$	image of a set, 16
f^{-1}	inverse map, 17
$g \circ f$	composite map, 17
$id_{\mathcal{M}}$	identity map, 17
$\mathcal{N}(x)$	equivalence class of x, 21
$\mathcal{M}/\sim$	quotient space, 22
ν	natural map, 22
$\mathcal{W}$	set of all words, 23
C^{-1}	inverse curve, 34
$C_1 \cdot C_2$	concatenation of C_1 and C_2, 35
$W(D)$	set of all curves in D, 35
$C_1 \sim C_2$	equivalence (homotopy) of curves, 35
$W(D;O)$	set of all closed curves in D with base point O, 40
$\pi_1(D;O)$	fundamental group of D with base point O, 41
$[C]$	homotopy class of C, 41
$\approx$	homeomorphism of spaces, 48
$f\vert_U$	restriction of a map to a subset, 54
f_*	induced map on fundamental groups, 59

$\deg(f), [D' : D]$	degree of a covering, 65	
$\Gamma(D' \xrightarrow{f} D)$	covering transformation group, 71	
$\sigma(P_1; P_2)$	covering transformation that takes P_1 to P_2, 72	
$\tilde{P}$	point with tail, 78	
$\tilde{D}$	set of all points with tail, 78	
$V(D; O)$	set of all curves in D starting at O, 79	
$e(C)$	endpoint of C, 80	
$U_\epsilon(P)$	disc of radius ϵ and center P, 81	
$\overrightarrow{PQ}$	line segment from P to Q, 81	
$\tilde{U}_\epsilon(\tilde{P})$	ϵ-neighborhood in $\tilde{D}$, 81	
$(D'; O') \xrightarrow{f} (D; O)$	$D' \xrightarrow{f} D$ is a covering and $f(O') = O$, 83	
$\approx_\Gamma$	equivalent by a covering transformation, 83	
$\tilde{D}/\Gamma, \Gamma\backslash\tilde{D}$	quotient space for $\approx_\Gamma$, 83	
$C^0(D)$	set of all continuous functions on D, 89	
f^*	induced map on functions, 90	
$\mathrm{Aut}(C^0(D'))$	group of automorphisms of $C^0(D')$, 91	
$C^0(D')^\Gamma$	set of all Γ-invariant functions on D', 91	
$O(D)$	ring of holomorphic functions on D, 93	
$\int_C F(z)dz$	line integral of F over C, 93	
$K(D)$	field of meromorphic functions on D, 94	
$z(P)$	complex coordinate function of P, 94	
$\dfrac{\partial F}{\partial G}$	derivative of a function with respect to another function, 95	
$O(D')^\Gamma$	set of Γ-invariant holomorphic functions on D', 98	
$V_\#$	solution space of the differential equation (#), 105	
$\dim_{\mathbf{C}}$	complex dimension, 106	
$\mathcal{M}(\gamma)$	monodromy representation at γ, 108	
$[w_1, w_2]$	basis of a vector space, 108	
$\mathrm{GL}(2, \mathbf{C})$	group of invertible complex 2×2 matrices, 108	
Σ	a set of known functions, 109	
$L_0(\Sigma)$	functions of type L_0 on Σ, 110	
$L(\Sigma)$	functions of type L on Σ, 110	
$U(a, \epsilon)$	open disc of center a, radius ϵ in $\mathbf{C}$, 114	
$\tilde{U}(a, \epsilon)$	spiral staricase covering $U(a, \epsilon)$, 114	
$\arg(\tilde{p})$	argument of $\tilde{p}$, 114	
$r(\tilde{p}; a)$	modulus of $\tilde{p}$, 114	
$\log$	logarithm function on a covering space, 115	
$(z - a)^\alpha$	exponentiation on a covering space, 115	
c.p.s.	convergent power series, 122	
$B(\tilde{D})$	set of meromorphic functions on $\tilde{D}$ with regular singular points, 12	
$M(\tilde{D})$	field of quotients of $B(\tilde{D})$, 129	

$\mathbf{C}(z)$ field of rational functions on $\mathbf{C}$, 129
R Riemann sphere, 99, 129
$\lambda_i, \mu_i, \lambda_\infty, \mu_\infty$ roots of indicial equation, 132

Index

Γ-invariant, 91

5-yen coin, 114

accumulation point, 94

anti-homomorphism, 91

argument, 114

automorphism, 91, 99

binary relation, 20

Cauchy's Integration Formula, 3

Cauchy's Theorem, 3, 94

Cauchy-Riemann equations, 3, 93

class, 19

closed curve, 33

composite map, 17

concatenation of curves, 35

concatenation of words, 24

conjugate curves, 70

conjugate points, 70

connected, 35

continuous map, 3

contour integral, 3

copiable neighborhood, 61

copy, 61

covering class, 85

covering map, 55

covering space, 55

covering transformation group, 71

covering transformation, 71

degree of a covering, 65

Einstein, Albert, 22

equivalence relation, 21

equivalent coverings, 84

equivalent words, 25

first homotopy group, 44

free group, 29

Fuchsian type differential equation, 3, 119, 130

function of type L, 110

function of type L_0, 110

fundamental group, 44

fundamental transformation of words, 25

Galois covering, 70

generator, 29

group, 3, 29

holomorphic function, 3, 93

homeomorphic, 48

homeomorphism, 3

homogeneous linear differential equation, 105

homomorphism, 3

homotopic, 36

homotopy class, 41

identity map, 17

image, 13, 16

indicial equation at ∞, 132

indicial equation, 125

initial conditions, 105

initial point, 33

injection, injective, 16

integral domain, 94

inverse map, 17

isomorphism, 3

Jordan canonical form, 121

Laurent series, 3, 97

length of word, 23

lift, 64

linear transformation, 4

manifold, 78

map, 13

meromorphic function, 3, 94

modulus, 114

monodromy representation, 108

Morera's Theorem, 3, 94

natural map, 22

neighborhood, 3

normal covering, 70

normal subgroup, 3

null-homotopic, 42

one-to-one, 16

onto, 16

open set, 3

orientation, 33

partition, 19

Poincaré group, 44

point with tail, 78

polar coordinates, 115

pole, 3, 94

power series, 3, 94

Presley, Elvis, 22

principal part, 97

process of type L, 110

process of type L_0, 109

product of curves, 35

product of words, 24

projection of a curve, 61

quotient field, 97

quotient space, quotient set, 22

rectifiable, 46

regular singular point at ∞, 129

regular singular point, 117, 129

representation of a group, 4

restriction of a map, 54

Riemann sphere, 99

ring endomorphism, 91

ring homomorphism, 907

ring, 89, 93

shadow, 65

simply connected, 44

spiral staircase, 79, 114, 129

Stolz sector, 115

subgroup, 3

subring, 89

surjection, surjective, 16

terminal point, 33

torus, tori, 49

triangulable representation, 108

universal covering surface, 78

word, 23

Wronskian, 120

zero divisor, 94